Residential Surveying Matters and Building Terminology

This is an ideal reference book for students (undergraduates and postgraduates) studying Building Surveying, Quantity Surveying, or Architecture, etc. It should also be of use to the Construction-related legal profession, Property Managers and Letting Agents. Builders (and homeowners, interested in identifying faults in their property), should also benefit from this book.

Residential Surveying Matters and Building Terminology covers a wide range of new and old building terms, techniques, technologies and materials, but much more extensively than the average dictionary. The alphabetical format makes it easy to check up on terms and subject-areas quickly – and the detailed coverage (including helpful drawings by the author) provides clear guidance to the reader.

This book covers a multitude of subject-areas, including condensation problems, cellar rot, wet rot and dry rot, thermal cracks, settlement cracks, metal wall-tie corrosion-and-expansion cracks, subsidence cracks, roof-spread recognition, bulging- and/or leaning-walls, etc. Further subject areas include inspecting and analysing residential building-structures, both internally and externally; appraising underground drainage systems; and personal commentary on survey report writing.

Les Goring is a member of the Chartered Institute of Building (CIOB), with a recent affiliation to the Royal Institution of Chartered Surveyors (RICS). He was a Senior Lecturer at Hastings College of Arts & Technology, having previously been a lecturer at South East London Technical College. In addition to being a Member of the Institute of Wood Science, he is also a Fellow of the Institute of Carpenters and a Licentiate of the City & Guilds of London Institute. Since 2007 until recently, he carried out Building Surveys in a jointly-partnered company named Residential Building Surveys Ltd.

Residential Surveying Matters and Building Terminology

In Alphabetical Order

Les Goring

Routledge
Taylor & Francis Group

LONDON AND NEW YORK

First published 2023
by Routledge
4 Park Square, Milton Park, Abingdon, Oxon OX14 4RN

and by Routledge
605 Third Avenue, New York, NY 10158

Routledge is an imprint of the Taylor & Francis Group, an informa business

British Library Cataloguing-in-Publication Data
A catalogue record for this book is available from the British Library

Library of Congress Cataloging-in-Publication Data
Names: Goring, L. J., author.
Title: Residential surveying matters and building terminology :
in alphabetical order / Les Goring.
Description: Abingdon, Oxon ; New York, NY : Routledge, [2022] |
Includes bibliographical references. |
Summary: "Residential Surveying Matters and Building Terminology
covers a wide range of new and old building terms, techniques,
technologies, and materials, but much more extensively than the
average dictionary"– Provided by publisher.
Identifiers: LCCN 2022002843 (print) | LCCN 2022002844 (ebook) |
ISBN 9781032286716 (hbk) | ISBN 9781032253916 (pbk) |
ISBN 9781003297925 (ebk)
Subjects: LCSH: Building–Dictionaries. | Building–Terminology. |
House construction–Dictionaries. | House construction–Terminology. |
Building inspection–Dictionaries. | Building inspection–Terminology.
Classification: LCC TH9 .G67 2022 (print) | LCC TH9 (ebook) |
DDC 690.03–dc23/eng/20220411
LC record available at https://lccn.loc.gov/2022002843
LC ebook record available at https://lccn.loc.gov/2022002844

ISBN: 978-1-032-28671-6 (hbk)
ISBN: 978-1-032-25391-6 (pbk)
ISBN: 978-1-003-29792-5 (ebk)

DOI: 10.1201/9781003297925

Typeset in Times New Roman
by Newgen Publishing UK

To my kin named Goring, Eglington and Clee, and to the readers of books by me.

Contents

Preface   ix
Acknowledgements   x

A   1

B   15

C   43

D   76

E   93

F   99

G   121

H   130

I   140

J   144

K   147

L   149

M   161

N   171

O   174

P   178

Q   195

R   197

S   212

T   243

U   259

V   262

W   265

X   278

Y   278

Z   278

Preface

This book started out as a *Glossary of Building-Terms* in a six-page, A4 format of approx. 3000+ words and it was given freely to our small, surveying company's clients to assist them (if need be) in their reading and understanding of such terms as *fascia-* or *barge-boards*, *lead flashing*, *flaunching* and *window-reveals*, etc., likely to be used in the detailed reports that I/we wrote after surveying their under-offer property purchases.

The small, two-man company, named Residential Building Surveys Ltd (RBS), was under the Directorship of myself and Mark Kenward – and was regulated by the Royal Institution of Chartered Surveyors (RICS). But sadly, in recent times, RBS had to cease trading. My colleague, Mark Kenward (who had a second string to his bow) is still trading as Osbourn White Ltd., carrying out surveying work all over the South East of England – but I retired to my study to expand the RBS *Glossary of Building-Terms* into this commercially-sized book, covering *residential building-surveying matters.*

The book, of course, still covers building-terminology, but much more extensively; and the alphabetical format makes it easy to check up on new or old terms quickly. But terminology now takes a much lower place in relation to the main coverage of understanding *building- and surveying-matters* related to such subjects as material- and building-defects, recognizing and differentiating between cellar rot, wet rot and dry rot, or between thermal cracks and settlement cracks; metal wall-tie corrosion/expansion cracks, or subsidence cracks; inspecting and analysing building-structures externally and internally – including underground drainage systems – and personal comments on survey report writing, etc.

Target-readership wise, inspiration was triggered when RBS employed a talented part-time graduate-surveyor, one Sam Clee, MSc (Hons), a few years ago – and he knew of a graduate who had seemingly failed a job interview because of not knowing the modular size of a building brick. Thereafter, graduate surveyors became my target readers. Other professionals, such as Quantity Surveyors, Solicitors (Lawyers), Property Managers, Estate Agents and Letting Agents, might also welcome explanatory references to confusing terminology. Homeowners, of course, might also wish to increase their understanding and maintenance-knowledge of such valuable, yet vulnerable assets as built-structures.

Les Goring, AMCIOB, FIOC, FTCB, LCGI, MIWSc
Mark Kenward, MRICS, MCIOB, Mbeng, MIOC

Acknowledgements

The author would like to thank the following people and companies for their co-operation in supplying technical literature or other contributions used or referred to in this book. Special thanks go to **Mark Kenward,** a chartered surveyor, for editing and advising on the draft manuscript; **Darren Eglington,** my youngest daughter Jenny's partner, for his invaluable IT assistance; **Roy Robinson**, a life-long friend, for liaising with me regarding his detailed knowledge of leaded lights; **Peter Oldfield**, another helpful friend and colleague; **Peter Crittall** and **Sarah Waters** of Crittall Windows Installation Services (www.crittall.co.uk); **Steven Leach**, Technical Director and **Julie Stone**, Administrator, of Mould Growth Consultants Ltd (www.mgcltd.co.uk); **Rachael Ramsden** of British Gypsum Ltd; **Justin Peckham**, Head of Sales, ®Accoya wood UK, Ireland, Baltics, Portugal and India, for Accsys Technologies; **Kirsty Carr**, Marketing Executive and **John Heseltine**, of the technical team at Leviat®, a CRH Company of Helifix Sustainable Solutions (www.helifix.co.uk); **Chris Boon** of Poulton Remedial Services Ltd. (enquiries@poultonremedialservices.co.uk); **John Gomez** and **Lee Stewart**, of **Westerham Drainage Ltd** (www.westerhamdrainage.co.uk); **Phil Steele,** Sales Director, and **Bella Kaye** (Receptionist) of **RSM Lining Supplies (Global) Ltd**; and last, but not least, **Yvette Killian**, Customer Systems Manager for Taylor & Francis Book Royalties, for quickly resolving a receipt-of-payment problem, of royalties earned on my other books.

A

Accoya® wood: This trade description refers to a modern, revolutionary process used in transforming fast-grown pines (commonly referred to as **whitewood**, which previously had a limited industrial use) from well-managed, controlled forests, into so-called Accoya wood with greatly-improved durability and stability; thus, allowing such timber to be used in more demanding applications, for which it was previously unsuitable. It is claimed to be virtually rot-proof and insect-proof, with a durability greater than teak and it is warranted to last 50 years above ground and 25 years in the ground. This is achieved by the use of a modification process which involves immersion in acetic anhydride, which results in almost total exclusion of water molecules from the timber's cell walls. This happens as a result of changes to the wood molecules, that would otherwise draw in water from the capillaries. This so-called *acetylation process* is environmentally friendly and not just *skin*-deep; it permeates through the whole cellular structure and enables Accoya wood to be used for making exterior joinery items such as windows and doors and carpentry applications such as structural framing and cladding, etc. It can also be used with long-lasting results in wet-and-dry conditions when in contact with the ground. Exclusion of water from the cell walls of the processed wood, results in unparalleled stability in the face of changes in humidity, meaning that Accoya is far less likely to shrink, swell, twist, warp or cup, than any unmodified timber. One of Accoya wood's identifiable characteristics – brought about via the acetylation process – is that a very faint smell of vinegar is produced during the machining process.

Acid rain: Sheet lead used to cover flat roofs is initially a bright silver colour, but once exposed to the atmosphere, it changes to a dull grey via a coating of lead carbonate which forms naturally on the surface. This carbonate acts to protect the lead from deterioration, but future problems can occur from an attack of *acid rain*. This is caused when rain absorbs an excess of pollutants in the atmosphere in industrial areas burning large amounts of fuel, which can chemically affect the rain and cause damage to parts of the lead roof-surface.

ACM PE (Aluminium composite material, with an unmodified **polyethylene** core): This is the technical description given to the cladding material used on the exterior façades of the high-rise apartment-block, known as Grenfell Tower, London, that suffered a life-destroying inferno in 2017. It is also referred to as *ACM category 3*. Subsequent enquiries, research and tests (set up by the British Research Establishment – BRE) have concluded that ACM PE presented the most significant fire hazard in relation to tests carried out on similar cladding materials. It is believed that no conclusions have yet been reached regarding culpability.

Acoustic construction: In habitable buildings, this refers to walls, ceilings and floors, etc., that are designed or upgraded to reduce (or eliminate) sound leaving or entering a room or building – which is particularly critical (and difficult to achieve) in buildings that have been converted into blocks of flats/apartments. Although, if the conversion took place after the more stringent Building Regulations were introduced in 1992, regarding HMOs (Houses in Multiple Occupation), noise from neighbouring flats should be far less. Sound reduction, measured in decibels (dB), is usually achieved – in its basic form – by *discontinuous construction* methods, involving strategically placed insulation membranes and non-vibratory cushioning pads (acoustic underlay). A few basic examples are: double-skin, insulated walls and partitions; suspended, insulated floors with separate, insulated ceiling-constructions below; suspended, timber-joisted single floors (as in converted flats), with their joist-voids fitted with slabs of 100mm/4 in-thick rigid insulation – and with floating floors fitted above the original boarded deck; secondary glazing units (apart from any double glazing), with soft, insulation linings fixed to the windows' encased-soffits, window-board surfaces and side-reveals.

Regarding *houses converted into flats*, mentioned above, British Gypsum have now produced acoustic floor systems, referred to as GypFloor Silent, which are specified in their downloadable, big White Book. One of the systems is detailed for residential conversion or improvement work, whereby existing timber-joisted floors can be upgraded to meet the more stringent Building Regulations' requirements for separating floors between converted dwellings.

Additive or Admixture: A liquid substance added in small quantities to various fluid mixes of mortar, concrete, or rendering, etc., to alter its workability/plasticity/drying-time or for achieving a greater degree of waterproofing.

Aggregate: Stones and stone-chippings, gravel, slag-waste, sand or similar inert materials, that form the main ingredients of concrete, plaster, mortar, etc. Aggregate is determined as *coarse aggregate* if it stays on the wire-screen of a sieve with 5mm square holes and as *fine aggregate* (such as *coarse* sand) if it passes through them.

Air bricks/grilles: Clay (or cast-iron), perforated grilles, with a stretcher-brick-face size of 220mm x 65mm/8⅝ in x 2⅝ in, which were/are built into walls to ventilate the underside of floors – mainly when such floors were/are constructed of timber joists. Nowadays, such air grilles are usually made of dull-red plastic. Ideally, vents placed across the front elevation of a dwelling (just above DPC-level and at about 900mm/3 ft. centres), should line up with those placed across the rear elevation (to promote better cross-ventilation). Traditionally (prior to modern-day condensation problems), air bricks sized 220mm wide x 220mm high were built at high-level into the walls of habitable rooms – especially bedrooms. And although this gave necessary ventilation (in such high condensation-producing rooms), it did not address the low room-temperatures in properties that already had poor heating systems, such as a single, open coal-fed fire. Such problems were not solved until the mid-20th century, when it was discovered that a balanced combination of improved *insulation*, controlled *heating* and *ventilation* were required.

Air conditioning: A mechanical system designed to bring purified air into a building or dwelling, at a desired temperature and humidity.

Air duct: An enclosed channel of square, rectangular, or tubular sectional-shape, usually pre-designed in thin, galvanized sheet-metal, plastic, etc., to act as a confined passageway for extracting foul- or moisture-laden air from a bathroom/WC/kitchen, etc., via an extractor fan.

Airing-cupboard shrinkage phenomena: Traditional airing cupboards, housing a hot-water storage cylinder positioned on the floor, or just above the floor, with an arrangement of slatted shelves above for airing laundered items, are often found to be the cause of excessive shrinkage in the surrounding building-elements. Close inspection usually reveals that the floorboards and floor joists have shrunk excessively in that localized area – sometimes causing longitudinal shrinkage-cracks in the ceiling/wall or coving/cornice junctions of the rooms below. This floor-depth shrinkage caused by the confinement of the hot-water cylinder – also often affects one of the cupboard's door-lining legs, causing it to drop – and in turn, causing the top of the airing-cupboard door to bind (jam up). It may also cause damage to the junctions of the ceiling/wall or coving/cornices of the rooms at the cupboard's level.

Air lock: In plumbing, this is a stoppage or interruption of water-flow in a pipe and is due to bubbles of air being trapped in it – detected by the loss or disturbance of water-supply at the tap. Air locks are only likely to occur in systems with low-pressure, cold-water storage cisterns and are triggered by badly-run horizontal pipework, which should be either dead level, or have a slight fall to allow air to escape from the system upon filling. For comparison, see **Water Hammer**, covered under that heading in this book.

Air-seasoning of timber: Historically, once trees for commercial use had been felled, then cut into transportable boles (logs) and eventually sawn into basic sectional-sizes, the process of reducing their moisture-content had to be started. This was essential, because unseasoned (green) timber holds too much moisture in its cellular structure to be used commercially. So, seasoning was carried out via storage in timber yards or open-ended sheds, after the stacked-up timbers had been separated with *piling sticks* to allow air to circulate between them. This time-consuming process, which varied from a number of months to a few years (for certain hardwood species), was eventually superseded by time-saving kiln-seasoning. And, although many die-hards still prefer/preferred air-dried seasoning, it must be mentioned that it tends to react badly in centrally-heated buildings, compared with controlled kiln-seasoned timber, that can achieve the required lower moisture-content of below 20% mc.

Air-sourced heat pump: See **Heat pump**

Air test: This was a traditional plumbing method of testing the hand-made cement-joints of glazed earthenware drainage systems, by using screw-plugs to seal the entries and/or exits of drainage-runs, pumping air in and testing the pressure with a U-gauge connected to the highest screw-plug. If the U-gauge detected any small leaks in the new pipe-run's joints, a Building Inspector would not give permission to back-fill the drainage excavations.

Alcove: An indented recess – also referred to as a *niche* – in a wall; historically used in sitting rooms, dining rooms, or hallways, as arched recesses – with or without shelves.

Algae: A group of plants that contain chlorophyll, but do not have stems, roots or leaves. One form of algae is single-celled and forms moss-like growths on roofs and damp walls, etc.

Alligatoring/crocodiling: A paint-film that has broken-up (crazed), displaying a pattern thought to be similar to a reptile's skin. Usually caused by exterior paintwork having been applied to the substrate during freezing or near-freezing temperatures, causing a lack of adhesion between the surface-moistened undercoat and the viscous finishing coat.

Alternating-tread stair: *Figure 1*:   A stair with a steep pitch (about 60^0), constructed of paddle-shaped treads, with the wide, usable portion alternating from one side to the other on consecutive treads; as per Approved Document (AD) K1, of the Building Regulations. Note that, although the stair's steepness is allowed to exceed the usual Regulations, the staggered, alternating treads allow the 2R+G rule to be retained.

Having experienced ascending and descending such a stair, one realizes that they ought to be used like a ladder, i.e., by using both handrails and climbing up and down backwards. Attempting to descend in a frontward-looking position is extremely scary and is personally believed to be potentially very dangerous.

Aluminium windows and doors, with double-glazed sealed units: These were the fore-runner of uPVC windows and doors and – although the aluminium is obscured by a white-plastic coating – they can still be seen around in a number of properties. Without close inspection, they are usually distinguishable by their slim, unmoulded

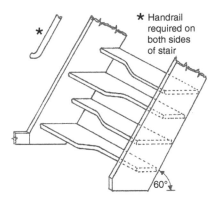

* Handrail required on both sides of stair

60°

Figure 1 Paddle-shaped alternating-tread stair

(non-bevelled) sections. Also, they are usually set within rebated, timber subframes, comprised of hardwood.

Anaglypta wallpaper: Very thick, plain wallpaper with heavily embossed patterns, which was very popular in the Victorian era for lining the lower-half of walls below dado rails. The anaglypta was usually stained-and-varnished – presumably to protect the surfaces from damage and/or to increase longevity.

Angle closers (usually just referred to as closers): Short lengths of brick, cut to one-quarter of a brick's length (or half of its width), to form (*close*) the bond at the corner (quoin) of a brick wall being built in Flemish- or English-bond.

Angle/dragon-ties: *Figure 2*: These traditional, hip-base timber constructions – being historically related to the use of *dragon beams* – were used in 18th- to 19th-century properties, on large *site-cut*, pitched roofs, to combat the tensile-stresses on the hips. This was achieved by anchoring the hips' stub-tenoned bases to the corners of the building via non-projecting *dragon beams*, attached to timber *angle ties*. The ties were fixed diagonally across the wall-plate corner-junctions to tie them together. These time-consuming constructions were superseded by simpler timber- and metal-ties (as illustrated) in the early 20th century.

Annex: An attached or separate part of a building that provides an extra facility, such as, for example, a *granny annex*, or an *office annex*.

Anti-capillary grooves: *Figure 3*: These are small, semi-circular grooves that are formed around the side-edges of casement-window sashes and frames, to inhibit rainwater from creeping in via capillary attraction. They should also be formed on the front, underside

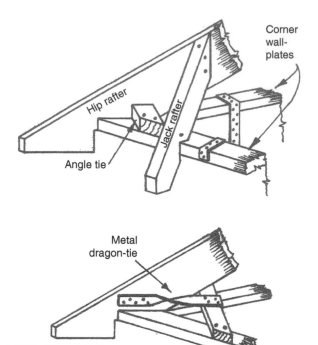

Corner wall-plates

Hip rafter

Jack rafter

Angle tie

Metal dragon-tie

Figure 2 Angle and Dragon-ties

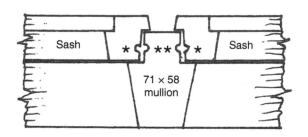

Sash

* ** *

Sash

71 × 58 mullion

Horizontal section through a BWMA casement window showing ★ Anti-capillary grooves

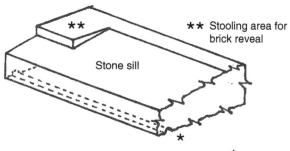

** Stooling area for brick reveal

Stone sill

Anti-capillary groove (throating) ★

Figure 3 Anti-capillary grooves

edges of window-sill projections where they are often referred to as **drip grooves** or **throating** grooves. As well as being semi-circular shaped, they may also be quarter-circular (quadrant-shaped) or right-angled (square, trench-shaped). But semi-circular are more common nowadays.

Anti-siphon pipe: Traditionally, this is a pipe which admits air to the discharge side of water seals of a number of traps jointed to a common (single) waste pipe; thereby ventilating the system and (critically) preventing the water seals being sucked out (siphoned) when water is released from any single or multiple outlets in a system.

Anti-siphon trap: A waste trap that is designed with less risk of losing its critical *seal*, by increasing the volume of retained water in the trap, or by enlarging the lower part of the U-trap. Such traps in a single-stack system of plumbing can save the expense of an anti-siphon pipe.

Appraising low-rise buildings: See **Structural appraisal of low-rise buildings**

Approved Building (Regulations') Inspectors: See **Building Control**

Approved Documents (AD):   See **Building Regulations**

Apron flashing: A simple, right-angled piece of *sheet lead* (or other material) *flashing*, used below vulnerable timber window-sills, or on *one* side of a chimney stack where its juncture with the roof forms an obtuse and/or acute angle. See also **Stepped-flashing**.

Apron linings: *Figure 4:*   Planed softwood or hardwood boards (or other material, such as plywood or MDF) that provide a finish to the vertical faces of sawn-timber surfaces, such as the faces of trimmer- or trimming-joists of a stairwell.

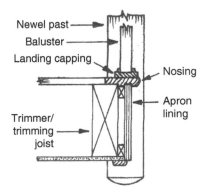

Figure 4 Apron lining fixed to landing-edge

Apron wall: This usually refers to the portion of wall beneath a window sill, which should be protected with a right-angled *apron flashing*, as described above.

Arboricultural Report: A detailed analysis and recommendation from an arboricultural specialist concerning a tree, or trees on a site – usually after some concern regarding the activity of a tree's roots. Such a report is sometimes recommended by a building surveyor, after he has completed a Building Survey on a property. Arboriculturists sometimes dig a small, but deep test hole on site, to ascertain the make-up of the subsoil.

Arcade: A fully-roofed, enclosed walkway area, usually glazed, with access to shops and stores on one, two or more sides.

Arch-bars/iron-supports: Historically, camber-shaped or flat iron bars, with a sectional size of about 63mm x 6mm/2½ in. x ¼ in., were used to permanently support low-rise segmental arches in common (rough) brickwork over fireplace-openings. Nowadays – with the possible renaissance of fireplaces – the bars and the arches would be superseded by precast, reinforced-concrete lintels.

Arch bricks (technically named voussoirs):   These are tapered (wedge-shaped traditionally by hand) bricks used to form traditional arch shapes. The tapered (in situ) brick-shapes and the underside (intrados) of such arches radiate from geometric centre points.

Arch centres: This terminology refers to temporary wooden structures (referred to simply as *centres*) upon which brick- or stone-arches are formed. The construction of the centre can be simple or complex, depending mainly on three factors: **1)** the span of the opening; **2)** whether its geometrical shape is complex; and **3)** how many times the centre is to be used for other identical arch constructions. For small spans up to about 1.2m/4 ft., the centre can be simple, of single-rib, twin-rib or four-rib construction. But for spans exceeding 1.2m/4 ft., the centre becomes more complex and, of multi-rib construction.

Arches as lintels: Nowadays, arches (if used at all) are usually for aesthetic, design reasons although there are a great number of arches of all shapes and sizes still in evidence in the UK, as a testament of their fitness to be used as load-bearing lintels. However, conversion work on old-build properties informs us that rough, segmental-arches in common brickwork over fireplace-openings, were used as lintels, to be plastered over. Of course, this construction was historically replaced by shallow-depth, precast reinforced-concrete *plank lintels*.

Architect's Certificate: On architect-controlled/quantity-surveyor-involved building contracts, once the work has reached certain stages of completion, signed certificates authorizing interim payment are usually issued to the contractor or Site Agent/Manager.

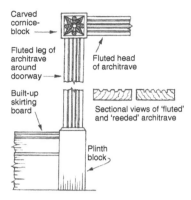

Figure 5 Architrave/Plinth blocks & built-up skirting to a corniced doorway

Architrave/plinth blocks: *Figure 5*:   The need for these may only now be necessary on refurbishment work in old buildings, where deep, built-up skirtings, often involving two or three stepped skirting-board sections (thereby increasing the skirting's thickness), could not be masked at the doorway by the relatively thinner architraves. The ends of the built-up skirting were housed into the sides of the plinth blocks by about 6mm/¼ in. to offset any shrinkage across the blocks and the base of the architrave legs were half-lap jointed and screwed into the back of the blocks, to make the blocks an integral part of the *legs*.

Architraves: See *Figure 5*: Plain or moulded sections of timber or other material, such as MDF (medium-density fibreboard), that are mitred and fixed around the face-edges of door openings, etc., to add a visual finish and – essentially – to cover the joint between the plastered- or plaster-boarded wall and the doorframe or door-lining.

Armoured cable: Electric cable for use on the exterior of buildings. In addition to ordinary insulation, this two-core or three-core cable is protected by a close-mesh steel-wire armour sheath, which is insulated with an outer sheath of PVC. When buried underground, though, such cables must be covered with cable covers or red-warning tape.

Armour-plated glass: As the name implies, this refers to better-quality, thicker glass that has been toughened (for safety reasons) to British Standards and CE EN 12150.

Arris/arrises: The sharp external edges (the dihedral angles) on timber or other materials. It has been traditionally established that all sharp arrises should be removed (planed or sanded off) on good quality joinery, or on certain items in second-fixing carpentry – such as door linings or doorframes and doors. Sharp arrises can cause paint-film peeling, or cuts to hands, etc. Their removal is also thought to improve the finished appearance of joinery or carpentry. Most of these considerations may also apply to other objects made of metal or plastic, etc., or to worktops made of granite, quartz/ polymer resin, etc.

Arris hip-tiles: These tiles – which are of clay or concrete – may be used with plain-tiles as an alternative to bonnet hip-tiles. Their positioning on the hips is similar, whereby they butt up against the sides of the plain tiles on each side of them and they are fixed with one nail at the top (which is overlaid by the subsequent tile above) but their main difference is that, because of their *squarish* (trapezoidal) shape and their range of ordering-options to suit the varying *dihedral planes* of different roof-pitches, the tiles do not need their bottom edges bedding in mortar.

Arris rails: Fence rails with a triangular sectional shape, usually with their ends scalloped to fit into pre-cut mortise holes in the 75mm x 75mm or 100mm x 100mm fence posts. The shape of the rails is partly for economy of material, i.e., one 75mm x 75mm sawn timber cut through diagonally to produce two arris rails and partly related to increasing the sectional strength of the rails, whereby the face-depth of the rail (being the hypotenuse of the triangular shape) has a depth of 105mm/4⅛ in. The top of the arris rail also forms a weathering-slope.

Arris (or Angle) ridge-tiles: Such tiles, manufactured in clay or concrete, are (as their name suggests) of an inverted, widely-spread V-shape. Their mortar-bedded attachment to the ridge is similar to other ridge tiles, but they are generally thought to be more visually appealing.

Art Deco buildings: This description refers to a style of building-design related to the 1920s and 1930s, which was characterized by a contradictory mixture of innovative geometric shapes and avant-garde colours *internally* – in contrast to the visually-simplistic structures *externally*. Usually, this meant that there were no visible pitched roofs (i.e., the roofs were shallow-sloped (technically flat) and lined (clad) with heavy-gauge sheet-lead, secreted by parapet walls. And the plain, white-stucco exterior walls and slim-framed, thin glazing-barred metal (Crittall) windows, added to these buildings' simplicity. One non-simplistic feature, though, was the shallow-projecting bay-windows, which were usually quadrant-shape ended (requiring expensive bow-shaped glass). The background to the Art Deco period is written herein under: **Art Nouveau buildings**.

Artex-coated ceilings or **walls:**   As these textures are formed with a thick paste made from Artex powder mixed with water, it must be understood that *prior to 1984/85*, potentially health-threatening *chrysotile asbestos fibres* (also known as *white asbestos fibres*), were used in the manufacture of Artex. Therefore, it seems likely that quantities of *shop-stocked*, original Artex (containing asbestos fibres) may have been sold and used in homes until these stored-stocks ran out, *after* the mid-1980s. However, although the manufacturers themselves took the initiative and stopped using this health-threatening material in 1984, its use was not officially banned in the UK until 24

November 1999 – although **amosite** and ***crocidolite***, the two most dangerous asbestos fibres – were banned in the UK in 1985.

Nevertheless, texture-coated ceilings and (to a much lesser extent) walls are in evidence in a great number of UK properties built between the mid-1950s and 1999, and, therefore, for health and safety reasons, the question of whether the coatings applied during this period contained chrysotile asbestos fibres, should always be raised by building-surveyors, knowledgeable builders and trade-knowledgeable homeowners.

The main industrial-reason for the proliferation of textured coatings, using patterns of *stippled*, *swirl*, *fan*-shape and *broken-leather* – apart from any aesthetic reasons – was that it was a cheaper alternative to the more time-consuming practice of traditional wet-plastering with a 3 to 5mm thickness of smooth finishing plaster. And, although Artex coatings have also been applied to *plastered* ceilings, in my surveying experience, such ceilings have always turned out to be of traditional lath-and-plaster, suggesting that they were texture-coated to cover up historic cracks and bulges commonly found in this type of old ceiling.

Textured coatings can be tested by specialists, to determine whether asbestos fibres exist. Alternatively, removal can be done by non-specialist decorators (or DIY individuals) by using an allegedly-safe solvent Eco solution stripper known as 'Home Strip X-TEX ARTEX REMOVER'. But it must be borne in mind that the process (involving the use of a decorator's *stripping knife*) will very likely damage the substrate surface of the un-plastered plasterboards – thereby requiring them to be *set* (skimmed) with finishing plaster. I have anecdotal knowledge that some plasterers, if requested to do so, will plaster over old, contaminated Artex coatings, regardless (or ignorant of) the current Health & Safety Regulations requiring such concealment to be marked with a warning notice.

Finally, it must be mentioned that some people take the view to not disturb their textured ceilings and to 'let sleeping dogs lie'. With such a view, it is believed that if the likely health-threatening material is not damaged or disturbed (scraped, cut, broken, drilled, or deteriorated with age), there is no health risk.

Artificial stonework: Good-quality imitation stonework is made in reusable mould boxes, traditionally made from close-grained timber such as Obeche; they are pre-cast under factory conditions to produce artificial-stone components such as window sills, lintels, arches (in voussoir-segments), doorway jamb- and ashlar quoin-blocks, coping stones, etc. The oiled mould-boxes are lined with a *stiff*, liquid-mix of crushed stone, selected sand and white cement. This mix (of about 25mm thickness) is placed and tamped carefully to all surfaces of the mould box; then a semi-stiff mix of concrete is placed and tamped into the *stone*-lined mould. Usually (for reasons of tensile-reinforcement and/or transportation-stress), steel reinforcing rods are placed carefully (and strategically) into the concrete during this casting operation.

Art Nouveau buildings: This description refers to an avant-garde building-design that started in the late 1890s, but was superseded by *Art Deco buildings* in the early 1900s. Art Nouveau started the movement away from over-ornamentation of buildings, by using stark, simplistic structures with stucco-rendered and painted surfaces, innovative geometric shapes and very low rise, depressed semi-elliptical arches. Their plain, interior walls were relieved with plant-life motifs. For comparison, see **Art Deco buildings**, under that heading above.

Asbestos-cement roof slates: This description refers to pre-WWII slates that were made from asbestos fibres and Portland cement. They were mass-produced as a cheaper alternative to traditional slates and tiles made from quarried slate, stone, clay or concrete. They carried a brand name of 'Eternit', inferring that they would last for an *eternity*, but they lost favour in the mid-20th century when the health-hazards connected with asbestos fibres became an issue. Two Eternit slate-designs existed, which could be obtained in a variety of colours, ranging from grey (their natural colour), red, blue or russet-brown. The first of these two designs were called *straight-cover* slates – and as their name implies, they were rectangular-shaped replicas of traditional quarried slates. The second design was called ***diagonal-cover slating*** – the slates of which were square-shaped and laid to form a 45^0 saw-tooth pattern; which was either referred to as ***diamond-cover slating***, if the lower points of the diagonal slates were left intact – or referred to as ***honeycomb*** effect slating, if the lower points of the diagonal slates were cut off horizontally.

It must be mentioned that Eternit-type ***straight-cover*** cementitious slates are still in use, but are now minus the asbestos fibres, since the industrial use of such hazardous material has been banned in the UK since 1985. The problem, therefore, that Surveyors have on Building-surveys, is not only determining the condition of a pitched roof's covering, but of also determining its age – if possible.

Asbestos-containing materials (ACM):   Large quantities of raw asbestos fibre, mined in Canada and Southern Africa throughout the 20th century, were imported into the UK and, because of its unique chemical composition and physical properties, it was originally hailed as 'the magic mineral'. It could withstand fierce heat, had soft and flexible fibres that were resistant to alkaline attack and it was considered indestructible and able to resist decay under most conditions. These characteristics made the most industrially-used of these fibres (***chrysotile***, also known as ***white asbestos***), a useful reinforcing component in the prolific manufacture of asbestos-cement sheet material and many other building products.

However, for many years now, it has been known that the magic mineral had a devilish downside: it was discovered to be a carcinogenic health hazard, causing diseases such as *mesothelioma*, *lung cancer*, *asbestosis* and *laryngeal cancer*, etc., and a ban on its use has been extremely slow. In the UK, the final ban on chrysotile, commonly thought to be the least dangerous of the three main asbestos fibres, was not officially made until August 1999. And although this date was ahead of an EU directive banning its use, chrysotile asbestos fibres were actually removed by some companies (one being the manufacturers of Artex, the textured ceiling-coating) in 1985; this also being the date when the use of the other two widely-used asbestos fibres (amosite and crocidolite) were banned. Amosite is also referred to as *brown asbestos* and crocidolite (thought to be the most lethal of the three) is referred to as *blue asbestos*.

Although, in the UK, chrysotile fibres were known to have been used in building products to a far greater extent than amosite and crocidolite, it must be understood that some products used a mix of two or three of the fibres, making it impossible (without specialized testing) to be certain of which materials contained one or both of the most lethal asbestos fibres – and understandable when site-surveyors report their findings as being 'a *likely* ACM' (asbestos-containing material). With this in mind, the following list of construction materials and/or components (in no particular order) are known to have contained one or more of the three health-hazardous fibres, in either ***residential or industrial properties:***

1) Textured coatings, such as Artex, predominantly on ceilings, occasionally on walls;
2) Asbestos insulating boards and or asbestos-cement panels or tiles, used on ceilings, panels and walls, etc.;
3) Asbestos-cement soffit boards to eaves' projections, or undercloak-tile support-soffits at gable-wall verges;
4) Asbestos-cement corrugated sheets used predominantly on the roofs of garages and low-rise industrial outbuildings and sheds;
5) Plain asbestos-cement sheets and asbestos insulating boards used in garages for ceilings, walls and as facings for FR (fire-resisting) doors or used to replace ceilings in dwellings, after WWII bombing-raid explosions caused the collapse, or partial collapse of many lath-and-plaster ceilings. Note that such *temporarily repaired* ceilings still exist in a number of pre-war dwellings and one should be suspicious when they appear to be (or are heralded as being) *panelled ceilings*, because asbestos-sheeting was often panelled uniformly with cover-moulds. This was partly done to conceal any untidy side-abutment joints, but also to create a more pleasing, uniform appearance. So, they were covered, either with plain, unmoulded timber-fillets (of an ex 50mm x 12mm/2 in. x ½ in. section), or, more realistically like a panelled ceiling, with astragal-moulded sections, scribed at their intersections.
6) Asbestos-cement roofing slates for pitched roofs, either rectangular likenesses of quarried slates, or diagonal slates with 'diamond'-shaped points that formed zig-zag patterns across the roof's surfaces. They were manufactured in natural grey but blue, red, russet-brown and crushed, natural-slate-surfaced asbestos-cement slates were also available;
7) Bitumen felt and tar products used on flat roofs and bitumen felt used as sarking-felt over the rafters of pitched roofs;
8) Thermoplastic vinyl floor tiles – and the bitumen used to adhere them to the sub-strate (which also acted as a DPM (damp-proof membrane));
9) Fibrous weatherboarding and profiled wall-cladding panels;
10) Promenade tiles, used to create a walkway on flat roofs – and their bitumen adhesive;
11) Sprayed asbestos that was used in industrial buildings for heat- and sound-proofing – and as a protection against fire and condensation on steel beams and stanchions. Apparently, at least two asbestos fibres were used in this mixture – and it must be borne in mind that sprayed asbestos is extremely friable unless sealed;
12) Loose asbestos lagging used for heat- and soundproofing, loft insulation and insu-lation between ceiling- and floor-joists;
13) Preformed moulded products, such as black plastic WC cisterns and toilet seats, white-asbestos water-storage tanks (often found disconnected and redundant in loft-voids), sewer pipes, rainwater guttering and downpipes, 110mm Ø soil-and-vent pipes (SVPs) and their branch-pipes serving WCs, wash-hand basins and sinks;
14) Asbestos cement flue pipes to boilers;

15) Loose-fill insulation (such as vermiculite granules), traditionally laid between ceiling joists in lofts;
16) Vermiculite granules mixed with cement and water and laid to a 50mm thickness as insulation below sand-and-cement floor screeds laid on concrete substrates;
17) Laggings on boilers and pipes;
18) Asbestos insulation board (such as Asbestolux) to FR (fire-resisting) doors;
19) Asbestos insulation (cast in situ, like plaster-casting) around boilers, storage heaters and warm-air heating systems;
20) Asbestos insulating board used as bath panels, fixed behind fuse-boxes, fitted fires, and to the inner faces of boiler/airing-cupboard doors;
21) Asbestos-rope seals (gaskets), as historically used on AGA cooker flue-pipe joints;
22) *Old*, unused fire-blankets are known to be very health-hazardous if used, because when they are pulled out of their housing, dangerous asbestos-fibres are immediately released into the air. *Modern* fire-blankets are, of course, safe to use.

Ashlar/ashlaring: Oblong-shaped hewn stones laid to a stretcher-bond pattern to form a stone wall; Also, this terminology may be a reference to the *appearance* of ashlaring formed on the face of sand-and-cement rendered and painted (stucco) walls.

Ashlar quoins: *Figure 6*:   The corners (quoins) of a building formed with alternating header/stretcher ashlar blocks, as a decorative relief to the otherwise plain-rendered or brick-walls. Such quoins may be formed with sand-and-cement render, quarried stone, or precast artificial stone, terracotta, etc.

Ashler stud walls: *Figure 7*:   Vertical, timber-stud partition walls fixed in attic rooms from floor to rafters, to partition off the lower, acute angles of the pitched-roof slopes.

Asphalt roofing: See **Mastic asphalt**

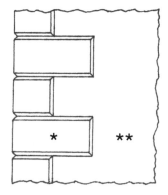

Figure 6 *Ashlar quoin to **rendered wall

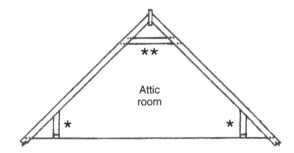

Figure 7 *Ashlar stud walls and **ceiling collars

Asphalt skirting: When asphalted, flat roofs usually meet up against parapet walls, etc., where an asphalt skirting – usually of two brick-courses high (150mm/6 in.) is required to combat lateral penetrating-dampness. And, to increase the skirting's limited edge-connection to the flat roof, a 45⁰ asphalt-angle-fillet of 50mm/2 in. width, is usually formed at the base of the skirting, soon after the skirting has been formed.

Atrium: **1)** An inner courtyard – usually with open- or gated-access – surrounded by a building (or buildings), a dwelling (or dwellings) on all four sides, thus providing more daylight via the additional courtyard windows; **2)** An entrance hall (vestibule), especially one that incorporates an upper area; **3)** Nowadays, the term Atrium is being used to refer to the glazed, isosceles-shaped triangular area of a gable-ended window, usually above an entrance or hallway.

Attic: See **Loft**

Axed bricks: Arch bricks that are cut roughly (by hand with a bolster-chisel or scutch) to a voussoir-shape that radiates to a common centre.

Axle pulley: See **Sash pulley-wheels**

B

Back boiler: A traditional domestic boiler fitted out of sight (at the rear) of a fireplace hearth, behind an open, solid-fuel fire, or connected to a gas or electric fire; the back-boiler is fitted with high and low threaded-connectors to receive primary flow and return pipes.

Back-filling or **back-fill:**   A mixture of hard rubble, broken bricks or stone, etc., and 20 to 30% of the excavated earth, all used to refill the sides of foundation trenches, after traditional strip-foundations have been laid and the walls have been built up to (at least) DPC level.

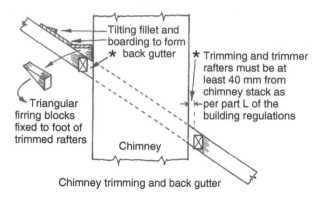

Figure 8 Chimney trimming and back gutter

Back gutter or **chimney gutter:** *Figure 8:*   Such gutters are formed at one side of a chimney stack where it passes through and forms an acute angle to a pitched roof. Usually – in these rainwater-trapped areas – wedge-shaped timber bearers (known as *firring blocks*) are fixed to the trimmer-rafter/trimmed rafters and a gutter-board is nailed into position. Then a second chimney-width board abuts this edgewise and is fixed to the trimmed rafters; finally, a *tilting fillet* is fixed to this board's top edge. Prior to tiling or slating, the back gutter is lined with lead, or other impervious material. Note that more sophisticated back gutters can be built with two-way firring pieces, to allow tapered (firred) gutter-boards to form a central ridge, allowing the gutter to 'fall' each way from the chimney stack's rear-centre.

Backing up: This term is used to describe signs of a possible drain-blockage, when liquid effluent is discovered lying motionless in an open drainage-channel of an inspection chamber (manhole).

Bakelite: One of the earliest forms of plastic, produced in the early 1900s; it was brown-coloured and was used for door-knobs, electric-light switches, etc. On building-surveys of old-build properties, Bakelite switches are occasionally still found to be in use; such being a sure sign that the electrical wiring needs to be recommended to be checked out by a qualified electrician/electrical engineer.

Balanced construction: See **Plywood**

Balanced flue: This refers to the design of a room-sealed gas- or oil-appliance that draws its combustion air in, within the same flue pipe that discharges its combustion products to the outside of a building, usually via the assistance of a fan. Note that, for safety reasons, if the exterior balanced-flue terminal is to be fitted less than 2m above ground/footpath level, it must be fitted with a wire-mesh guard. Also, to prevent flue-gases re-entering a building, the flue-terminal's position must be taken into account in relation to nearby windows and doors.

Balanced steps: See **Dancing steps**

Ballast: Apart from cement and water, ballast is the main constituent of concrete and is comprised of graded stones of certain sizes, such as 19mm sieve-size, and 'sharp' sand of a certain size that has to pass through a 6mm-mesh sieve. Another much-used mix of concrete uses stones that have passed through a 9mm-mesh sieve.

Ball cock: This is a float-operated valve in a WC cistern, that controls the amount of flushing-water released to supply water to a WC (water closet); the spherical ball – traditionally of copper, but nowadays of plastic – is attached to a metal or plastic arm, controlled by a lever handle.

Balusters: See *Figure 4* on page 7: Also called **spindles**. These are usually of lathe-turned, slim wooden posts – but can be square or of other material, such as metal – fixed between the handrail and string-capping, or the handrail and the landing-edge, as part of the balustrade (guarding) of a staircase or landing.

Baluster sticks: As described above, but only of square-section posts with no lathe-work.

Balustrade: The barrier or guarding at the open side of a staircase or landing, comprised of newel posts, handrail and balusters or of solid panelling, etc., topped with handrailing.

Balustrade height: Current Building Regulations' Approved Document (AD) K1, the Amended Edition, stipulates that handrail heights in all buildings should be between 900mm and 1m., measured vertically to the top of the handrail from the pitch line or the floor of a landing. Note that the *pitch line* depicts a theoretical acute-angle that touches all the front edges of a stair's nosings for handrail-heights, but is also used to check the application of the 2R+G formula against any tapered steps in a stair's flight.

Band-and-hook hinges: These are heavy-duty *strap hinges*, also referred to as hook-and-ride hinges. The bands are made of wrought-iron strip, the end-knuckles of which drop on to heavy-duty pins, which are either built-in to a brick pier, or fixed onto wooden jambs or posts. Traditionally used on framed, ledged, battened and braced garage-doors and gates.

Banister: This traditional term, rarely heard nowadays, means **balustrade**, i.e., a row of plain or lathe-turned balusters fitted and fixed below a handrail. However, the term banister was archaically also applied singularly in reference to a handrail.

Bare-faced tenon(s):   This term refers to tenons with only one shoulder, as is commonly used on stair-strings with only a 28mm finished thickness. The omission of one shoulder allows the bare-faced tenons to be of a more realistic thickness of 16mm, allowing the tread- and riser-housings to be of 12mm depth without clashing with the inner-face of the tenons.

Bare, full, or tight measurements: 1) A *bare* measurement (in Building terminology) means slightly *under* the quoted or actual size of a component; **2)** A *full* measurement means slightly *over* the quoted or actual size, and **3)** A *tight* measurement means the exact, quoted size. Note that in taking measurements for ordering glass for windows or doors, etc., the tight sizes must always be reduced by at least 3mm/⅛ in. to allow for any irregularities and for thermal movement.

Barge boards: *Figure 9*:   Fascia boards fixed to the uppermost faces of inclined (raking) gable-end walls, or (traditionally) to the ends of an extended ridge board, extended purlins and extended wall-plates (that bridge over the gable wall), immediately below the projecting roof-verges. Barge boards support the projecting roof-verges and act as an architrave, to mask the junction of different materials. Traditionally, barge boards were moulded or pierced with ornate fretwork. In modern-day roofing, plain-edged barge boards are fixed to so-called *gable-ladders*, which are preformed units, resembling two wooden ladders. The outer edges of the 'ladders' are fixed to the face-sides of the first trussed-rafter, or (in traditional roofing) they are fixed to the face-sides of the first pair of common rafters.

Barrel bolt: As the name suggests, this type of bolt – commonly used at the high and low lock-side edges of a door – is encased within a barrel. The barrels are either of a

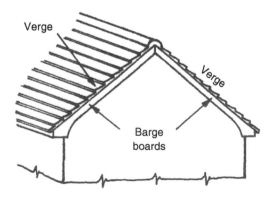

Figure 9 Barge boards and roof-verges

square section or round and are fitted with square or round sliding bolts. They can be obtained in a variety of metal finishes and sizes.

Baron's bend: An SGW (salt-glazed earthenware) open-topped bend used in a traditional inspection chamber (called a *manhole* in the recent pre-pc era). The open-topped mouth of the short-length bend is embedded in sculpted concrete-and-mortar called *benching*, up against the edge of the inverted SGW half-round open-channel, that carries foul water and effluent en route to the public sewer or septic tank, etc.

Basement rooms and/or cellars: Historically – and even occasionally nowadays – areas of habitable buildings built below ground, with one or more parts of their structural, outer walls up against the exterior ground, nearly always have or develop dampness-problems. This is because they are wholly or partly *subterranean* and are therefore subject to rising- and lateral-damp penetration. Even if such walls and concrete floors have been *tanked* by 3-layer asphalt-sandwiching, problems of ground-water penetration can still arise – especially if a particular location is known to have a high-water table – and if the retaining brick-walls against the asphalt tanking were built shoddily. Known instances are on record of sunken pumps having to be installed in the concrete-and-asphalt-sandwiched floor, as a permanent feature, shortly after tanked areas had been completed. Therefore, the question should always be raised about the habitable-suitability of basement accommodation.

Bastard tuck-pointing: See *Figure 59(e)* on page 191: This traditional lime/putty pointing of the mortar-joints of face brickwork, whereby very narrow, protruding nibs of pointing were tediously formed to leave a slight projection from the face of the brickwork, gained its crude, archaic reference to illegitimacy by being different to more commonly splayed pointing, or, perhaps, because achieving the narrow projections reduced a bricklayer to swearing.

Bat: This bricklaying-term refers to a part of a cut-brick that is greater than a quarter.

Batten: This traditionally refers to an indiscriminate length of sawn or planed timber, but usually of a relatively small sectional size of 50mm x 25mm/2 in x 1 in, or of 75mm x 25mm/3 in x 1 in. It may be used as a measuring or gauging rod, or as a fixing batten, to hold a component in place whilst it is being built in, etc. It may also refer to specific uses, such as a *tiling-* or *slating*-batten.

Battered walls: These are *retaining walls* built to purposely lean over (leaning walls) to hold (retain) high-level ground on a sloping or split-level site. The hydrostatic pressures that can build up against such walls should be combatted by building-in a number of small-diameter, open-ended *weep-hole pipes* (equal in length to the wall's thickness) near the base of the wall, usually at 0.6m to 0.9m centres.

Bay windows: Such windows project outwards from their main walls (beyond the designated *building line*) and – unlike oriel windows, which are supported on brackets, etc. – their plan-view shape, whether segmental, semi-circular, semi-octagonal, or semi-rectangular, etc. becomes part of the habitable room and carries on down to the foundations. Unfortunately, though, such foundations (usually laid separately to the main foundations) were often skimped, as evidenced by common settlement/subsidence cracks found in numerous bays' brickwork in later years.

BBA: The **British Board** of **Agrément** was initially established by the British Government in 1966 as a quango, which is now an independent certification body, offering certification and inspection services regarding construction work and materials in the UK. Their BBA services, approvals and inspection schemes are well recognized by local authorities' Building Control Departments/Partnerships, architects and designers and insurance companies.

Bead/beading: This usually refers to a small semi-circular-edged softwood moulding, rectangular in shape, with a quirk-edge on one side of its rounded edge – traditionally used extensively around the inner edges of boxframe windows to retain the lower, sliding sash.

Beam-and-block floors: See *Figure 23* on page 58: This relatively modern construction is used at ground-floor level on dwelling houses. The factory-made, reinforced concrete floor-beams take their bearings on the inner-skin (leaf) of blockwork and must be at least 75mm/3 in. above ground/oversite level to the underside of the beams. They are spaced out to receive the lengthwise dimension of standard wall blocks sized 440mm/17¼ in. x 215mm/8½ in. x 100mm/4 in., laid edge to edge, resting on the beams' protruding bottom-edges. The gaps between the blocks and beams are filled with a brushed-over sand-and-cement grout. The void beneath the precast beams should be provided with ventilation, continuous through any intermediate honeycombed sleeper walls, which may be running at right-angles to the beams.

   Once the building is weathertight, thick rigid-foam insulation is laid over the beam-and-block floor and a polythene vapour barrier is carefully laid over this, ready to receive a tongue-and-groove-panelled floating floor.

Bearers: Timber battens, usually ex 50mm x 25mm/ 2 in. x 1 in., that support shelves, etc.

Bed, bedding or bed-joint: A controlled thickness of mortar – usually 10mm/1cm/ ⅜ in. – beneath timber plates, bricks or blocks, etc.

Belfast/butler's sink: This traditional white-glazed earthenware, deep kitchen-sink, has been redeemed from its relegation as a plant-growing garden-utensil and has come back

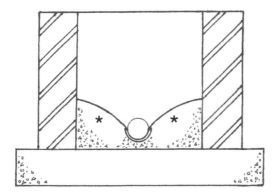

Figure 10 *Benching on each side of an earthenware channel in a traditional inspection chamber

into kitchen-use in recent years as being fashionable. Note, however, that another country seems to have audaciously changed its British nomenclature from Belfast to butler.

Bell-cast eaves: See **Sprocket-formed eaves**

Benching: *Figure 10*: This refers to the concrete-and-sharp-sand formation *sculpted* and formed around and above traditional glazed-earthenware channels and open-bends found in brick-built manholes – now referred to as *inspection chambers*. The benching helps to guide the effluent-discharge into the open, half-round channel, en route to the enclosed underground pipework leading to the sewage system – and by so doing, unwanted deposits are not normally left behind.

Bibcock: A water tap with a horizontal outer (male) thread at its inlet, allowing it to screw directly into a self-supporting fitting. These taps, fitted with a threaded nozzle for a hosepipe connection, are commonly used externally for gardening and/or car washing.

Bidet *(pronounced: Beeday)*: This sanitary appliance, designed to wash the unmention-able excretory regions of our bodies, does not appear to be generally popular in the UK. Either that, or designers of our box-like dwellings cannot squeeze one of them into their miniscule bathroom/WCs. However, note that bidets with an ascending spray must not be fitted to a mains supply; they can only be fed independently from a cold-water storage cistern. Also, the hot-water supply must be run independently from a storage-vessel. But, bidets with an over-the-rim type spray, via a pillar tap, can be fitted to the mains supply.

Bi-fold doors: Such doors are comprised of two halves that are hinged together on a vertical, central joint and operate in a common doorway-opening of 762mm/2 ft. 6 in. x 1.982mm/6 ft. 6 in. An inverted U-shaped channel is fixed to the underside-head of the door lining, in which castor-wheels, attached to the door-heads, enable the doors to run and fold. At the folding-side, a pin-pivot plate is fixed to the floor. Multi bi-fold concertina-style doors – in popular use nowadays, in a fully-glazed form at the rear of properties with gardens – operate in a similar, but more sophisticated way.

Bifurcated staircase: A staircase that stands independently of the walls (usually in the centre of a large hallway), and rises up to a two-way, quarter-turn landing, i.e., one can turn left or turn right (through 90⁰) to leave (or enter) the head of the stair.

Bill of Quantities: This is a document produced by a Quantity Surveyor, which gives a list of numbered, detailed tasks to be undertaken by a nominated building contractor. Each task/item details the quantity of work to be undertaken – and these detailed items are separately priced by a selected number of contractors to form a detailed offer/tender to do the work.

Binder(s): *Figure 11*:   In traditional roofing, binders are usually of 100mm x 50mm/4 in. x 2in. timbers fixed on their edges across the tops of – and at right-angles to – the ceiling joists. The fixings may be made by skew-nailing or with modern framing anchors. This is to *bind* the two components together, counteracting any deflection or mid-span movement in the ceiling joists if the span of the roof exceeds 2.5m/8ft. 2½ in.

Birdsmouth joint: See *Figure 11*: A traditional vee-shaped notch in timber, that was thought to represent a bird's mouth in appearance; traditionally found at the foot of a roof's common rafters, to give them a partial horizontal-seating on the edges of the roof's timber wall-plates, whilst allowing the top two-thirds of rafter to carry on down to form the eaves' projection.

Bituminous roofing felt: This traditional, water-resistant roofing felt, available in rolls of 1m width, is usually manufactured with hessian-mesh fibres running through it to

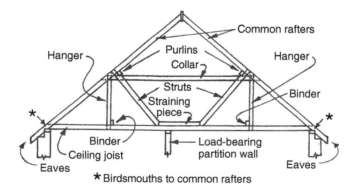

Figure 11 Traditional roof terminology

add extra strength when draped across open-rafters as a protective underlay. However, nowadays, to avoid limited roof-ventilation (traditionally provided at the eaves), **polyethylene breathable membranes**, such as *Tyvec Supro* or *Marley Supro*, are widely used, thereby achieving better (widely distributed) ventilation.

Bituminous roofing felt (without hessian-mesh reinforcement) may also be used on flat roofs, in built-up layers (usually three), bonded together with tarmac pitch. The top layer is/was traditionally finished with pitch-bonded spa chippings, but nowadays a mineral-finished top layer is used. These top-layer coatings add a degree of protection to these relatively soft roof-surfaces and increase their lifespan to a textbook expectancy of ten to fifteen years. However, there are thicker, mineral-felts nowadays, that are *torched*-on (with gas-filled blow lamps) to the felt underlays – and these have a guaranteed lifespan of twenty-five years. Details of these are given in this book at **Torched-on roofing felt**.

Black-ash mortar: This refers to a mortar-mixture containing ashes and lime. It was used in the late 19th and early 20th centuries for coloured pointing, but its main use seems to have been for cheapness in some counties of the UK. Such mortar is usually of very low strength. Nowadays, coloured mortars – if required for aesthetic reasons, such as *pointing* – are achieved by adding a coloured ochre pigment to the sand-and-cement mortar mix.

Black mould on walls and ceilings: See **Condensation problems in dwellings**

Black pitch-fibre drain pipes: See **Pitch-fibre drain pipes**

Bleaching: In painting and decorating, this term refers to the loss of colour, usually caused by a chemical action known as oxidation, brought about by sunlight.

Bleeder pipes: Small-diameter, wall-thickness pipes that are built into – and at the base of – retaining walls, to allow an escape route for trapped moisture in the high ground pressed up against such walls, which otherwise might exert structural-damage via hydrostatic pressures at their base.

Bleeding: In painting and decorating, this term refers to the unwelcome chemical-disturbance of a base colour that *bleeds* through to the colour recently applied above.

Blind hinges: Unseen, concealed hinges, such as *Soss hinges*, secreted (mortised) into the unseen edges of the door and frame or lining. Traditionally, such hinges were used on secret panels and/or doors to secreted rooms. One example of such work, [the construction of which was witnessed by the author] in a company boardroom in Mayfair, London, many years ago, was a secreted door within a façade of bookcase-shelving that covered the whole of one wall. The mid-area door – that concealed a small boardroom kitchenette – had been made with false, shallow-depth shelves projecting from its face, which lined up with the real book-shelves on each side. Once the false shelves had been fully loaded with shallow-depth false bookends, the illusion was complete.

Blind mortise: In mortise-and-tenon work – usually for aesthetic reasons – this refers to a mortise hole that does not pass wholly through the timber, because it is intended to receive a so-called *stub* tenon.

Blind nailing: See **Secret-nailed flooring**

Blistering: This term refers to small bubbles within a paint-film surface, usually caused by atmospheric moisture under the surface of the paint film.

Blockboard: This so-called *sheet* material is obtainable in 18mm/¾ in. thicknesses (which is thought to be the most popular thickness) and 25mm/1 in. thicknesses. The most common superficial size of these boards is 2. 4m x 1. 2m. The blockboard's side-by-side core strips are not-more-than 25mm/1 in. wide and run the length of the sheets, as does the grain-direction of the 3mm/⅛ in. thick face-plies.

Block-bonding: See **Toothing and block-bonding**

Block flooring: See **Parquet flooring**

Block partitions: This refers to partition walls usually built of aerated, foam insulation blocks measuring 440mm long x 215mm high x 100mm thick. Blocks can also be of 75mm thickness, but are more difficult to use on single-wall construction, because such thin walls tend to collapse if built too fast – unless side-supporting profiles are used – and/or the walls are built up in two or three stages, after the mortar has partially set and the lower blockwork is less likely to buckle.

Block or brick paving/pavers: In recent decades this historic practice of paving roads and footpaths has re-emerged in the form of paved driveways to privately-owned dwellings. Hard-faced bricks can be used, but coloured concrete-pavers seem to be more popular. A 200mm/8 in. to 250mm/10 in. depth has to be prepared below the required finished-surface, to receive a substrate of well-compacted Mot type 1 hogging, topped by a 25mm/1 in. to 38mm/1½ in. evenly-laid layer of sharp (coarse) sand upon which the pavers are carefully laid. Final jointing is done with kiln-dried sand.

Block plan: On new-build drawings, the basic outline of the building or buildings (*block plans*) are often shown on construction drawings at a scale of 1:1250, to identify the actual location and size of the proposed building's outline in relation to surrounding properties and named roads. The latter details are usually copied from Ordnance Survey maps.

Blockwork: This term refers to walls built with building-blocks, such as partition walls and inner-leaves (inner-skins) of cavity-wall construction, etc. The face-sizes of blocks usually approximate to two *stretcher-bricks* in length x three brick-courses in height, thereby allowing for incremental brick-and-blockwork courses (as in cavity walls) to correspond laterally and vertically. Bearing in mind the manufacturing-allowances made to accommodate mortar-joints to achieve *coordinated wall-sizes*, the face-sizes of building blocks are 450mm x 225mm/18 in. x 9 in. Blocks vary in thickness from 75mm/ 3 in. to 225mm/9 in. The latter thickness being commonly used for load-bearing walls, and the thinner blocks of 75mm and 100mm (especially the latter) being commonly used for partition walls. Present-day blocks, with high thermal values, are commonly used for the inner-skins of cavity-wall construction. And, of course, another reason for coordinated heights of six bricks and two blocks equalling each other, is so that a necessary line of essential wall-ties can be bedded in the mortar bed-joints of such incremental heights for cavity-wall construction.

Blown: **1)** In glazing, this is a reference to double-glazed sealed units that – because of their age – have lost their air seal and are 'weeping' internally with condensation: an obvious sign that the units need to be replaced. **2)** Blown is also a building reference to an area of plastered/rendered wall-surface that has lost its bonding-adhesion to its sub-strate, either partly or wholly. Some blown surfaces – especially small areas on internal walls – are best left alone if visually non-detectable, but others may need to be carefully *hacked off* and re-plastered or re-rendered. Note that if there is no visual evidence of bulges, cracks or more-serious surface-damage – most blown areas of rendered walls go undetected and are only usually discovered by surveyors on surveys, if their suspicions lead them to gently tapping (*sounding*) the surfaces randomly.

Blue bricks: See **Staffordshire blues**

Blue stain/sap stain: A blueish fungal discolouration sometimes found in sap-wood – seen in the form of short, irregular streaks in some sapwood-areas of par (planed all round) softwood boards – which is scientifically reckoned to be benign and non-destructive.

Blushing: In decorating, this term relates to a milky appearance of a lacquered sur-face, which can be caused by moisture in the paint or on the substrate.

Boiler: A water heater which – contrary to its terminology – does not (or should not, for domestic reasons) bring water to *boiling* point. They are usually heated by gas,

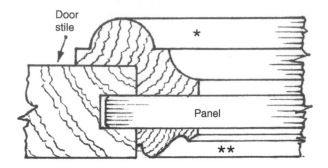

Figure 12 *Bolection mould and **Grecian-ogee panel mould

electric, or oil – although this may change, if the powers-that-be jointly tackle carbon emissions – and are thermostatically controlled. They need regular servicing and maintenance by qualified and certificated engineers/tradespeople. And this non-compulsory ritual is strongly recommended to be carried out annually. However, annual, certificated-servicing of boilers is compulsory for landlords under Building Regulations, when properties are tenanted. The landlord must organize the servicing and provide the tenant(s) with a copy of the detailed certificate.

Bole: This usually refers to the substantial part of the tree trunk – minus its branches – that is ready to be converted (sawn) into saleable timber.

Bolection mouldings: *Figure 12*:   Traditional, rebated mouldings fitted around inset, wooden door-panels to achieve an extra degree of embellishment. Such mouldings should be skew-nailed to the stiles-and-rails-only, to allow for thermal movement of the panels; or, on exterior doors, the mouldings should be slot-screwed through the panels from the interior side, before the interior, 'planted' moulds or bolection moulds are fitted and skew-nailed. Bolection mouldings might also be used to accommodate a glazed panel in a door.

Bomb damage: Although WWII bomb damage to dwelling houses and other buildings, etc. occurred during the early to middle years between 1939 and 1945, such a period of time is not really that long ago to preclude evidence of non-repaired damage still being on display to an enquiring eye. Of course, apart from towns like Liverpool, it might only be detected in London and other fringe areas usually, where the bombers disposed of their bombs en route back to base. Three of these unlucky dumping grounds were Hastings, St. Leonards-on-Sea and Bexhill-on-Sea, in the coastal area of Sussex. And, during Building-Survey inspections in these areas in recent years, three such dwellings (with no glaringly obvious signs of bomb damage) were personally appraised and likely bomb damage was reported upon.
    The discovery of such damage evolves from having developed an eye for straight lines and verticality – and being curious about slightly inward-bowed and outward-bowed

walls, in relation to acquired knowledge of damage to buildings caused by explosions and implosions. Mostly, such damage presents itself in the form of outward- or dogleg-bowed walls between ground-level and eaves' level of two- or three-storied dwellings. On one such incidence, subsequent inspection in the roof void showed that the force involved had moved one side of the pitched roof away from its connection to the ridge board (via the mid-area common-rafter plumb cuts) by about 38mm/1½ in. The roof appeared to have been re-slated, but no attempt had ever been made to repair the disconnected topside of the skeletal roof, or the partially-disconnected purlin-struts (although, other amateurish-looking struts had been inserted). Other internal evidence was revealed at the right-angled junctions of floor- and skirting-boards, where the mid-area wall-movement had left the outer edges of the floorboards uncovered by the faces of the skirting boards.

Although the weekly *Hastings & St Leonards Observer* newspaper published a detailed map of the bombing-incidents many years ago – which might still be available from their archives – it can also be seen at 1066online history, via www.1066online.com., along with details of the hundreds of high-explosive bombs dropped and the number of deaths and injuries. Personally, having done numerous surveys in the affected areas, an old copy of the *Hastings & St Leonards Observer*'s published map, served me very well in confirming any evidence found that suggested bomb damage.

Bomber butts: See Helical hinges

Bond: See *Figures 38, 42* and *68* on pages 96, 110 and 237: 1) In brickwork, these terms refer to long-established named patterns (such as *English bond*; *Flemish bond*; *Stretcher bond*, etc.) created partly for appearance, but mainly to give interlocking structural-strength to a wall. Such elevational patterns are formed by using the narrow ends (headers) of bricks and/or the long sides (stretchers) of bricks. **2)** The gluing and adhesion of timber components and joints, etc. **3)** The interface-adhesion of applied-plaster to walls or ceilings, etc. **4)** In roofing, the overlapping arrangement of various impervious-materials' perimeter edges, which also rely on roof-slopes/pitches to offset the risks of capillary attraction. **5)** In electrical wiring of a dwelling – especially of bathrooms and shower-rooms – this term refers to wiring or cabling that connects all metal components to each other and conducts/runs them to earth.

Bonding timbers: In a number of old properties, built prior to the early 1900s, horizontal bonding timbers, with a sawn, sectional size of 100mm x 75mm/4 in. x 3 in. were built into the interior-side of the exterior walls at recurring heights of about eight brick-courses, i.e., 600mm/2 ft., thereby replacing three or four rows of interior, common brick-courses. These bonding timbers (unseen when the walls were plastered) were used to strengthen the walls' thickness (of either 225mm/9 in. or 343mm/13½ in), but the effect of wet rot decay and dry rot destruction in untreated timbers such as these, buried beneath plastered surfaces in damp, single-leaf walls, seems not to have been known – or was disregarded.

In such aged properties, the existence of bonding timbers may only become known, or suspected, if – for whatever reason – a solid-plastered wall-surface (or a lath-and-plastered wall-surface) is hacked off for re-plastering; or excessive deterioration of the bonding-timbers has caused a degree of bulging to an outer-wall surface, or the inner-wall displays horizontal crack-damage to its plastered surface.

If deteriorated bonding timbers are discovered, repetitive hit-and-miss removal of them, in approx. 900mm/36 in. portions, and hit-and-miss replacement with common bricks is recommended. Note that this repair technique is similar to traditional underpinning-repair.

Bonnet hip-tiles: Half-cone shaped tiles (so-named because of their resemblance to a cloth bonnet historically worn by women and tied under the chin). They are of clay, or of concrete and are used on the dihedral angles of plain-tiled hips. They are bedded on their concave-shaped underside bottom edges (onto a top portion of the tile below) and are fixed at their top via one nail hole in the centre of the narrow, convex surface. This nailed fixing area is then covered by the mortar-bed of the next bedded-and-fixed tile being laid above.

Boot-lintels: Reinforced-concrete lintels, in vogue in the 1950s, that were used above doorways and windows in cavity walls; they were so-called because they were thought to resemble the shape of a boot in their cross-sectional appearance – whereby the outer-skin of face-brickwork was supported by the toe (for aesthetic reasons) and the inner skin of blockwork was supported by the deeper (beam-like) portion of the boot shape. Technically frowned upon nowadays for causing a phenomenon known as **cold bridging**; and potentially liable to suffer sooner or later from **concrete cancer**. Both subjects covered herein under the bold headings.

Borrowed-light windows: This is a reference to fixed (unopenable) windows built into internal partition-walls (usually in the upper regions only), for the purpose of borrowing natural daylight from exterior windows – as per BS 565.

Bottom-hung fanlights: These were used in traditional casement windows up until the mid-20th century and are, therefore, still in existence in numerous buildings. And, to create a weathered opening-arrangement, they were hinged on the inner face-edges of their bottom rails and opened (swung down) internally. This descent had to be limited, of course, so they were fitted with a pair of *quadrant stays* – one for each near-top side of the jambs' rebated edges. This restricted the opening-angle of the fanlight to about 60°, but to open it fully, the metal stays could be flexed/sprung slightly to each side. Finally, the fanlight sash was secured at the top by a ring-operated face-fixed latch and a face-fixed striking *keep*. Note that an essential feature of a bottom-hung fanlight, is that it should be fitted with a sloping weatherboard/drip-mould on its outer, bottom-rail face.

Boundaries: In building terms, these are important dividing lines between adjoining plots of land, that are initially outlined on Ordnance Survey maps – in all sorts of shapes and sizes. They are also shown in more detail on the original construction drawings of a property, to locate the outline of a building in relation to its surroundings. Usually, during the legal transaction of buying a property, a *block plan* outline view of the dwelling and its boundaries becomes available to the purchaser. The boundaries should be shown clearly on this, usually with the maximum length and breadth measurements indicated. There should also be a few boldly-marked letter T-shapes on the plot-owner's side of the fence, with the T's bottom-ends touching only those boundaries that are legally bound to be maintained by the plot-owner. There is a need for vigilance when purchasing a property, as many disputes arise from boundaries having been moved (even minimally) one way or another over time.

Boundary Survey Report: There are many disputes over the exact position of boundaries to a property, sometimes resulting in feuding situations between neighbours. It is therefore advisable – when purchasing – to check the land-registry block plan (usually received from a conveyancing solicitor during the early stages of purchase), to ascertain measurements of length and breadth shown thereon. If concerns regarding an erosion of plot-size arises after exchange of purchasing-contracts (or at any future date), the complainant is advised to engage a Building Surveyor to carry out a Boundary Survey. If such a Report is able to conclusively – yet tactfully – identify a boundary violation, such matters can sometimes be put right without litigation.

Bow: A segmental-shaped warp in the length of a board, springing from the wide face of the material.

Bow windows: These are segmental-shaped windows which project outwards from the main walls (beyond the legally designated *building line*), but – unlike bay windows, which carry on down to their own foundations – bow windows are usually carried by cantilevered supports or ornamental wooden gallows' brackets.

Boxed-heart: In timber-conversion mills, this term refers to a method of rip-sawing logs that are known to have defective *heart-shakes* (splits emanating from the central pith area of the heartwood), so that the defects are contained within a *boxed* length of timber, which may or may not be saleable. To produce the boxed heart, certain thicknesses of tangential-grained boards are cut from four sides of the log towards the estimated boxing.

Boxed shutters: See **Folding shutters**

Boxframe windows: *Figure 13***:**   These are traditional, wooden window-frames that incorporate side-boxes to house lead or cast-iron weights that counterbalance the vertically-sliding sash-windows. Nowadays, there are lookalike wooden boxframe windows with solid side-jambs and heads, whereby the glazed sashes are counterbalanced on patented *spiral balances* fixed to the side-jambs and housed into the sash stiles. There are also double-glazed uPVC units, made to look like boxframe windows, that are very realistic – and more energy-efficient.

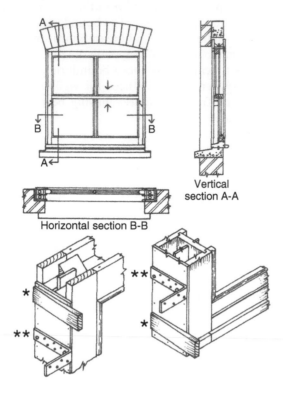

Vertical
section A-A

Horizontal section B-B

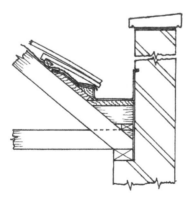

Figure 13 Box-frame window details *wedging positions and **metal fixing-anchor positions

Figure 14 Lead-lined Box gutter up against a parapet wall, capped with a weather-sloped
coping stone

Box gutters: *Figure 14*: Lead-lined box gutters are commonly used when a lead-
covered flat roof or the eaves of a pitched-roof (clad with tiles or slates) are so designed
as to be adjacent to a parapet wall. The box gutter, which runs parallel to (against) the
parapet wall, should not be less than 225mm/9 in. wide and it should be at least 50mm/2
in. deep at its highest end, when it is at the foot of a pitched roof (with its lower-portions

boarded); or, it should be at least 100mm/4 in. deep at its highest end, when used with lead-lined flat roofs. In the case of a box gutter serving the eaves of a pitched roof, the roof-side boarded-edge of the gutter-box must have a timber tilting-fillet fixed in place before the leadwork commences.

Brace: **1)** A diagonal support; **2)** A tool for holding and revolving a variety of drill bits.

Braced and/or trussed partitions: See **Trussed and/or diagonally braced partitions**

Brackets: **1)** Projecting-supports, usually made of metal or wood, and usually formed by a combination of horizontal members, vertical members and diagonal supports, forming so-called *gallows'* brackets; **2)** Short-length, splay-ended vertical-boards, usually with a 150mm x 25mm/6 in. x 1 in. sectional size, alternately side-fixed to the 100mm x 63mm/4 in. x 2½ in. rough-sawn undercarriage-supports of a geometrical staircase with a *cut-and-bracketed string*, to help support the underside width of the tread-boards.

Bracketed balusters: This refers to a limited number of protruding, wrought-iron stair-balusters, bent at right-angles at their base and built into the outer-string side of concrete or stone steps, or caulked with lead into preformed mortise holes in the stair's outer string.

Breach of covenants: See **Restrictive covenants**

Breathable-membrane underfelt: Water-resistant and breathable (vapour-permeable) fabric material, manufactured from high-density polyethylene. Available in 50m. rolls with 1m. or 1.5m. widths. Used nowadays as a roofing underlay in preference to traditional non-breathable, bituminous roofing-felt. Used on roofs and timber-framed walls, this modern innovation improves the energy-efficiency and thermal targets in a building by making direct ventilation unnecessary in both *cold-* and *warm-deck* constructions, whilst also providing an impervious under-cladding beneath the tiling or slating membranes.

Breeze blocks and **Breeze bricks:**   Pre-WWII building blocks and bricks, made from the ashes of coke or coal, mixed with sharp sand and cement. These blue-coloured, manufactured products, possessing aerated voids, were the forerunners of modern-day insulation blocks such as Thermalite, Celcon, etc., and they provided good *cut-clasp nail* fixings for second-fixing carpentry work. Breeze bricks – prior to modern-day instant-grab, panel-adhesives and wall-drilled screw-fixings – were often built-in to inner-skin *brick*-walls and *brick*-partition doorway-reveals to provide cut-clasp nail fixings for door linings and doorframes.

Bressummer or **Breastsummer:**   A large-sectioned, horizontal oak beam, traditionally used to support floor joists or brickwork above wide, inglenook fireplace openings – and also used to support ancient timber-studded/framed buildings, with brick-nog panelled façades, jettied-out from the recessed ground-floor elevation(s).

Brick and timber-framed cavity-wall dwellings: This description refers to one form of factory-prefabricated dwelling, where storey-height panels, consisting of timber-framed studwork, pre-clad with WBP plywood-cladding on their exterior face, doorway- and window-apertures where designed, are used as the inner leaf (skin) of the dwelling. Once the concrete oversite is formed and ready, the panels are erected and bolted together in their designed positions, temporarily braced and bolted down where necessary. And – if it was a chalet-bungalow, for example – the first-floor joists would then be laid and fixed across the shortest span. The prefabricated, plywood-sheathed gable-wall panels would then be fixed and temporarily braced, in readiness for the factory-made truss-rafter roof-assemblies. These would have a designed shape to suit the low-rise ashler walls and collar-tie-positioned ceiling joists of a loft-room – but be in the form of truss-rafter assemblies.
    Once this inner-shell of the building is erected, a continuous cavity is then formed by building an outer leaf (skin) of single-width facing-bricks – laid to a stretcher-bond pattern around the outside of the plywood-clad and sheathed panels. As this is being built, right-angle-shaped, non-corrosive metal, frame-cramp wall-ties are bedded on the brick outer-skin and screwed to the face of the ply-clad inner-skin panels – thereby creating the traditional horizontal and vertical criss-cross spacing arrangement of the essential tying of the single-skin inner walls to the single-skin outer walls. At the start of the second-fixing operation, prior to dry-lining, the timber-studded wall panels are filled with rigid-foam insulation material and clad with a vapour barrier.

Brick arches: Brick (or stone) arches over windows and doorways, in a wide variety of geometrical shapes, can only now be seen mostly on older-style buildings. Present-day design favours straight lines for various reasons, including visual simplicity, cost and structural requirements in relation to new materials and design. Curved arches in dwelling-house outer walls have been mostly replaced by various types of lightweight, galvanized, pressed-steel lintels. However, so-called brick soldier arches, displaying a row of face-side bricks built vertically on their ends, over windows and/or doorways (and as *string*-course features to a building's façade) seem to be still in vogue.

Brick-nogging to stud partition-walls: This was a yesteryear-practice that can still to be found in evidence in pre-20th-century properties. Apparently, it was a hybrid practice done to strengthen any single-skin internal walls of ½-brick-thickness, that were built with lime-mortar. After the rough-sawn 100 x 50mm stud partition was erected, which included diagonal braces, the awkward spaces in between the framework were filled with lime-mortar-bedded bricks. Of course, a similar construction can also be found on the façades of properties (some of which are classified with Grade I or Grade II listings), where single-skin *exterior* brick-nogged stud walls have been used. But, usually, the brickwork in-between these studs and braces can be seen to have been roughly rendered (plastered) and painted externally. Internally, such walls are usually found to be vertically battened and lined with wooden laths-and-plaster – thereby forming a cavity and achieving an early form of dry lining. However, the aged, untreated battens are likely to be affected by wet-rot decay.

Brick-on-edge copings and sills: See *Figure 71*: **1)** This refers to bricks bedded on their narrow edges across a wall's thickness to form a continuous line of level headers-on-edge, acting as a coping at the finished height of a garden-wall, etc. **2)** Window- and/or doorway-sills may also be formed in this way and may or may not have a weathered slope on their top surfaces – but should have a DPC (damp-proof course) in their horizontal bed joints.

Brick- or **block-wall terminology:**   Any brickwork or blockwork that is to have its mortar joints raked-out and 'pointed' (refilled with a stronger sand-and-cement-ratio mix) and trowelled to a neat finish, or have its joints pointed as the work proceeds (the present-day practice), is usually referred to as *facework*, *face-brickwork*, or *face-blockwork*, involving the use of *facing bricks*, or *fair-faced* blocks. And any brickwork not required to display a face-side (perhaps because it is to be rendered) is usually referred to as **common work**, or **common-brickwork**, involving the use of *common bricks* (or the **common faces** of *facing bricks*) – and their mortar joints are left in a rough, un-pointed state. And any common brickwork that is to be *roughly pointed* (and traditionally rubbed over with an old sack, or rough cloth, soon after construction), is referred to as having a fair-face finish.

Bricks: Building-bricks are usually made from clay or shale – known as *brick earth* – and they are either moulded by hand or machinery and burnt/fired (not baked) in a kiln. Clay suitable for brick-making must contain alumina and silica. Alumina makes the clay plastic and silica (usually in the form of sand) assists in producing hard, durable and uniform bricks. A wide variety of bricks are available, produced in many colours and textures. With the exception of so-called *engineering bricks*, most bricks are porous and, therefore, not waterproof as such; hence the progression from solid-wall construction to cavity-wall construction, whereby the outer leaf/skin gets wet, but dries out without the risk of affecting the inner leaf/skin. This phenomenon of *getting wet* and *drying out* can also be thought of as *being allowed to breathe* – a natural (and healthy) function which ought to be considered when applying dampness-inhibitor liquids to walls.

Brick-sizes: The standard *coordinating* imperial-size of a brick's length x depth x height, which includes a $0\frac{3}{8}$ in. (inch) mortar joint is/was 9 in. x $4\frac{1}{2}$ in. x 3 in. But, in practice, the *actual* size of an imperial-sized brick is $8\frac{5}{8}$ in. $\pm$ $\frac{1}{8}$ in. x $4\frac{3}{16}$ in. $\pm$ $\frac{1}{16}$ in. x $2\frac{5}{8}$ in. $\pm$ $\frac{1}{16}$ in. To make sense of this, it must be realized that the stated coordinates (that *include* the mortar joints) make it possible for designers to work out lengths (and heights) of brickwork to achieve linear arrangements that do not require cut-bricks.

The standard *coordinating* metric size of a brick, based on the above measurements, is 225mm x 112.5mm x 75mm (length x depth x height, which includes a 10mm mortar joint). So, the actual size of a metric brick is 215 mm x 102.5 mm x 65 mm $\pm$ a few millimetres.

It must be understood that when the UK switched over from imperial units of linear measurement to metric units soon after joining the EU in the early 1970s, only an approximation of the imperial units was made. The unit of one inch (1 in.), truly equalling 25.4 millimetres (mm) was de-fractionalized to equal *25mm* – and the imperial unit of one foot (1 ft), equalling 12 inches, became 12 x 25mm = *300mm* (instead of truly equalling 12 x 25.4mm = 304.8mm). Although de-fractionalizing was an understandably welcome move, the adverse implications of this in the Building Industry need to be understood, because metric-sized facing bricks do not marry-up to imperial-sized facing bricks on repair, maintenance and conversion works carried out on pre-EU old-build properties – which, is believed to outstrip new-builds in the UK – and such mixing of brick-sizes, which is reflected in the unavoidably larger mortar joints created, invariably looks unsightly.

Brick slips (Brick-tiles or **Brick-veneers):**   Whichever of these three terms are used by different manufacturers or individuals, brick slips are relatively thin slivers of realistic-looking artificial building bricks, which are adhered (like wall-tiles) to different substrates to create a false façade of solid brick walls. The one exception to the artificial material used – to my knowledge – is that a company named Reclaimed Brick-Tile Ltd., based in Manchester, use a wide range of reclaimed, good-condition bricks from which they cut their brick tiles.

Imitation brick-façades (and imitation stone-façades) have been around for many years now, having been used initially to change (or cover up) the exterior appearance of rendering. Nowadays, brick slips (more commonly referred to as tiles) seem to have re-emerged as a fashionable façade for interior wall surfaces. The outlets that sell brick slips usually also sell the adhesive and pointing mortar, the joint-spacers, and the pointing- and grouting-gun required.

Bridging of damp-proof courses (DPCs):   This term refers to the acceptable dampness in walls below the DPC, creating a problem when it superficially bridges over the DPC and thereby transfers dampness to the previously-dry walls above. Likely ways in which this can happen, are: **a)** Borders of gardening soil built-up against a wall, higher than the DPC at certain points; **b)** Sand-and-cement rendered plinths that have been added to the brick-faces below the DPC, but which have been formed (at least slightly) above the DPC – and often finished with porous-looking and rough, square edges, instead of having weathered, sloping edges; **c)** Brickwork façades that have been rendered and/or are of pebbledash and this add-on coating has not taken the DPC into account – and has bridged over it; **d)** Any attachment fixed to an exterior wall, that bridges the DPC; such as a gate post that soaks up rainwater. DPCs can also be bridged internally, by **e)** underfloor rubble and rubbish left by builders and occasionally found piled up against the inner-faces of outer walls, likely to cause bridging.

Bridging- or common-joists: See **Floor joists**

British Board of Agrément (BBA):   A British Government quango, originated in 1966 to promote innovative industrial ideas, products and practices in industry, by issuing

BBA certificates of appraisal and approval. The model was based on a similar Agrément Board which was set up in France in 1958.

British Research Establishment: See **Building Research Establishment**

British Standards Code of Practice: These are recommended good trade practices and standards for materials and components. In recent years, to comply (or concur) with EU standards, many Codes of Practice were given a BS EN number, such as BS EN 13598 for polypropylene inspection chambers. Whether the EN will be dropped, now that the UK has left the EU, is unknown at the time of writing.

British Standards Institution (BSI): An organization that standardizes the methods of testing and dimensioning of materials between the manufacturer and the consumer. They are also involved in the publications known as *Codes of Practice*.

British Standard Specifications (BSS): These are numbered publications of the **British Standards Institution**, which describe the required quality of a material, or the required dimensions of certain products like bricks, blocks, drainage pipes, etc. The recommendations in these publications are widely used by architects and surveyors, etc., who quote them in their project Specifications and Bills of Quantities, as abbreviated and numbered references, such as BS 336: 1995 (which covers sawn structural timber sizes, etc.).

Brown rot: See **Cellar rot**

Building-blocks: See **Blockwork**

Building Control: A reference to local Government Building Control officers and their department or to approved, private individuals and/or companies, authorized to oversee and approve building work which was once the exclusive domain of extremely experienced Building Inspectors, employed by a local authority. Although, no doubt, such *partners* – as they are now referred to – will be academically qualified.

Building Inspections: Buying a previously-owned and occupied building (house or flat, etc.), is not unlike buying a second-hand car. We are wooed by sales' personnel and shiny images – and may not be mindful of possible faults that may exist, unseen by viewing-purchasers. And if the seller/vendor (of a dwelling) knows about such faults or blemishes, they are not legally obliged to disclose them. So, after a satisfactory viewing and a successful offer has been made – regardless of not having seen anything wrong with a property (even if you have some building knowledge) – an independent building inspection is highly recommended. With new buildings, as with cars, it would not be unreasonable to expect them to be covered by a Warranty, such as that offered by the NHBC.

Building surveyors usually offer one or more of the following, at varying charges:

1) *Valuation Survey* (done by accredited RICS' valuation surveyors);
2) *RICS' Condition Report*;
3) *RICS' Homebuyer Report* (which has been recently changed and does not have to have a valuation included);
4) *Building Survey* (known as a Level-3 Survey for properties over 50 years of age);
5) *Structural Survey* (done by a Structural Engineer or an accredited RICS' surveyor).

Note that a RICS' Homebuyer Report is not suitable for dwellings older than two or three decades or for properties that have had alterations. And therein lies the problem of surveyor-diagnoses: aged properties (like aged people) usually have more ailments, which cannot be realistically discovered in a short period of time. So, I believe that surveyors (or their staff) should give some guidance to clients who might be requesting a quote for the wrong type of survey, based on the appeal of a significantly lower fee. If need be, surveyors need only look online at an Estate Agent's front-elevational image of the property in question, to be able to get some idea of age, related to its style.

Building Inspectors: Traditionally, such people, with good technical qualifications and broad-based practical experience in a major trade-discipline, were employed by local authorities to periodically inspect the quality and integrity of work on building projects during various stages of construction, such as site-excavation, foundations, DPC-laying, drain-laying, etc. But, nowadays, these responsibilities seem to be mostly (if not totally) farmed-out to private, licensed Inspectors with technical qualifications only, who are directly employed by builders and building-contractors, or by individual DIYers. Some countries, I believe, give their technical workers (surveyors and architects) a grounding in hands-on practical work – which seems more in balance with trade-workers who historically receive a grounding in technical education.

Building-line: This term is also referred to as the *frontage-line* and, as these terms suggest, it determines the outer wall-face of the building under construction. It is usually given on the site-plan as a measurement from the theoretical centre of an existing or proposed road in front of the proposed building. Building-line positions are set by the Local Authority and Ordnance Survey team, as being a strictly controlled distance to the face of a new building in relation to an adjacent road or roads. When a building is being built in between other dwellings, a line can usually be struck from their frontages.

Note that building-lines are a very serious subject and, historically, there are cases on record of partly-built buildings having violated these lines and being served with a Local Authority notice to demolish the wrongly-positioned structure. I know of one case in Erith, Kent and another – many years later – in Hastings, East Sussex.

Building paper: A semi-waterproof paper, usually in 1m-width rolls, used for retaining the moisture content in freshly-laid bays of in situ concrete, etc.

Building Regulations: Prior to the existence of Building Regulations, *Building- and Public-Health Acts* came into force in the late 17th century, after the Great Fire of London, and a multiplicity of *building by-laws* came into operation in different parts of the country, often in contradiction to each other. But in 1966, The Building Regulations were introduced to replace the by-laws and establish uniformity. Since then, they have been amended and reprinted a number of times, but they were generally found to be difficult to interpret by building professionals and builders, so, in the late 1970s, separate technical publications, referred to as **Approved Documents (AD)**, were introduced and each of them covers the various subject areas in the occasionally-revised Building Regulations. To my knowledge, many professionals and builders still find them too complex and difficult to interpret. A number of authors have written books to *try* to explain these complexities, but, with respect for their effort, the subject matter is very difficult to explain via rewriting.

Building Regulations Control and Approval: Understandably, given the opportunity, many of us might like to build what we like, where we like, without having to consider others – or without a need for authoritative control and approval – or disapproval. So, these restrictions are necessary and are managed by each Local Authority's *Control Partnership*. This was previously known as the local authority's *Control Department*, before a lot (if not all) of the site-inspection and building-control work was contracted out to technologically qualified, self-employed building Inspectors or partnerships.

All major or minor work being planned must be notified on a *Building Regulations Submission form*, obtainable from a local authority's Building Control Partnership. As all notifications are believed to attract fees nowadays, *Charge Scales* and *methods of calculation* are set out in Guidance Notes on Charges, which are available from LACP (Local Authority Control Partnerships) on request. You can also apply for full Building Control Consent on some projects.

Building Research Establishment (BRE):   This former Government body, previously known as the *British* **Research Establishment** and originally created in the early 1970s, was an amalgamation of four pre-existing quangos, i.e., **1)** The Forest Products Research Laboratory, **2)** The Building Research Station, **3)** The Scottish BRE Laboratory, and **4)** The Fire Research Station. Historically, this quango had an excellent reputation in the UK and internationally for its research and technical publications. But it was privatised in 1997 and is now owned by an organization known as the BRE (*Building* Research Establishment). But it is still known as the centre of building science in the UK and still has a high reputation.

Building Survey: See **Building Inspections**

Building Surveyors: Now more commonly referred to as *chartered surveyors*, this professional body of people (men *and* women nowadays) has developed over a number of centuries from skilled, trade artisans, to surveyors, architects and engineers. But the route now in the UK has lost its attachment to practical learning and is mainly via a university degree-course and professional membership of the RICS (Royal Institution of Chartered Surveyors) or the CIOB (Chartered Institute of Building). However, to partly redress the imbalance, graduates from university or colleges are usually treated as *graduate surveyors* until they gain some actual surveying and site-experience.

Building Survey Report: Such reports are mostly commissioned by prospective pur-chasers of properties other than new-builds. The builder of the latter usually indemni-fies a purchaser against defects for a certain period of time – but buyers of the former have to ascertain the condition of the property themselves. Hence, without the know-ledge or the experience to do this, the need to employ a Building Surveyor to carry out a survey of the property is most important.

Note that 'Building Surveys' (as opposed to Homebuyer Reports, reported on herein separately) are reminiscent of a bygone age when surveyors spent far longer on site, carried out drain-tests themselves – if they were suspicious of defects – and often used a builder-in-attendance to check out certain inaccessible elements via a ladder; or to assist in positioning (dropping) a heavy plumb bob to check out the verticality of suspicious *quoins* (corner walls). But, in reality, time overspent on site equals a quoted fee being an underpayment. And, whether using plumb bobs for verticality or modern-day red-beam lasers, the surveyor carrying out a very thorough survey has a massively responsible task to do in a relatively short period of time. Then, he or she has to assimilate all that they have seen (and noted), in relation to all that they suspect, form conclusions and write a detailed report without fear or favour – sometimes to the disappointment of the purchaser, the vendor and the Estate Agent.

Built-up door-linings: *Figure 15:*   Traditionally, internal doorway openings in public buildings, or large, private houses and the like, invariably had timber *door linings* that were fitted across the entire width of the doorway-opening's *sides* (technically called **reveals**). Such reveals that were only the sides of ½-brick-thick partition walls of 112mm/4½ in. thickness received door-linings that were, of course, similar to present-day linings (single-width boards) – but those with wide side-reveals of 1-brick, 1½-brick and 2-brick thicknesses (225mm/9 in., 340mm/13½ in. and 450mm/18 in.) received built-up door-linings.

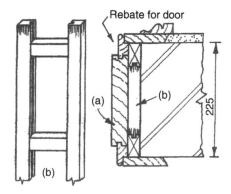

Figure 15 Built-up linings (a), fixed to framed-grounds (b)

Their purpose was to facilitate the hanging of a door, but also to protect the remaining width of the vulnerable parts of the otherwise-plastered reveals. And, to combat timber shrinkage over wide wall-reveal-widths, they were either comprised of two or three tongued-and-grooved lining-members fixed across the reveals' widths, or comprised of wide, built-up linings in the form of wall-panelling with sunken, beaded panels.

Modern-day door-linings, regardless of the width of the side-reveals, are positioned at one side or the other of the opening. And, disregarding the possibility of future damage, the unprotected side-reveals are either solid-plastered or dry-lined.

Built-up roofing: **1)** This description might refer to the traditional use of 3-layer built-up, bituminized roofing felt, seen ubiquitously on so many so-called *flat roofs* (which have slight *falls – gradients –* of up to 10^0); the felt is laid in three cross-layers, each bonded to its substrate with hot, brushed-on pitch. Traditionally, the top layer was also coated and then sprinkled profusely with white spa chippings, as a means of reflecting the heat from the sun.

2) The term Built-up roofing might also be applied to ***asphalt***, which is laid in two or three built-up layers on reinforced-concrete flat roofs, after a sand-and-cement screed has been laid to the required falls. Such roofs usually require an asphalt skirting to be formed at any parapet walls or other vertical-wall abutments – and these should be finished with top-bevelled (weathered) edges. Finally, the base of the skirtings should be finished with a 45^0 x 38mm/1½ in. wide asphalt fillet.

Built-up skirting boards: *Figure 16:* As mentioned here under the heading of *Architrave blocks or plinth blocks*, built-up skirting boards may only be encountered on refurbishment works in certain old buildings with storey-heights (floor-to-ceiling heights) far greater than those in present-day dwellings, which are usually about 2.44m/8 ft. So, *extra* wide skirtings used in *extra* high rooms, seem to be the design-reason for their origin.

Built-up skirtings are comprised of two or three tiered, stepped boards, with the top, moulded-edge board finishing against the plastered wall-surface. So, before the walls were plastered, a framework of *timber-grounds* (sawn or planed/prepared

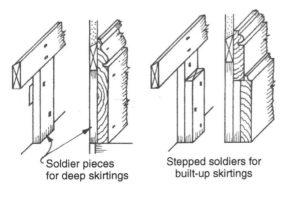

Soldier pieces
for deep skirtings

Stepped soldiers for
built-up skirtings

Figure 16 Built-up skirting boards

battens, with a bevelled top-edge to help retain the plaster-abutment) were fixed to the base of the walls to **a)** control and promote truer plastered surfaces and **b)** provide a stepped framework upon which to nail the two or three skirting members. The latter was achieved by fixing vertical *soldier* pieces (offcuts of ground) at 600mm/2 ft. to 900mm/3 ft. centres, below the top horizontal ground. In the case of a two- or three-tiered built-up skirting, these additional – shorter – soldier pieces would be required to be fixed to the first row of soldier pieces. The one or two longitudinal stepped-and-jointed connections of the skirting members were usually via a small groove and a cavetto-shaped projecting edge.

Bulging-wall phenomena: This may be visually more easily detected at the return-quoin (corner) of a building and may be related to rotational subsidence of a strip-foundation's width. If it is a slight bulge, it may not be detectable by eye, but rotational subsidence of a flank wall should be suspected if the appearance of a nearby vertical crack in *mid-area* of the return-elevation is detected. Such cracks have been seen to take place between the corners of a first-floor window-sill and a ground-floor window-head. But they may also occur between the corner of a first-floor window-sill and a ground-floor doorway-head vertically or diagonally below. However – confusingly – such cracks may also be due to ubiquitous thermal movement.

Also, when bulging walls are detected, one should question whether this might be caused via (a) wall-tie corrosion and expansion, or (b), ignorant removal of an interior cross-wall, inadvertently causing a violation of the *Slenderness ratio* related to wall-lengths and heights.

Finally, if significant bulging is detected to the exterior wall(s) of early- to late-19th-century properties, the decay of interior *bonding-timbers*, built horizontally into the walls, might be responsible. And could be confirmed by examining the interior wall-surfaces. For more information, see **Bonding-timbers**, written herein.

Bulkhead: This term has a number of related meanings, but in building, it refers to the *face-lined bulkhead trimmer* above the treads of a staircase. And the bottom, outer-edge arris of the bulkhead – to comply with Stair Regulations – must be not-less-than 2m vertically above the theoretical pitch-line of the stair.

Bullnose step: This is a quadrant-ended step at the bottom of a stair-flight that protrudes beyond the front face of a newel post. The protruding step is housed into the stair-string on one side and into the front-face of the newel post on the other. As well as being a decorative feature, the step's main, traditional purpose was to increase the stability of the newel post, by virtue of the outer-string's tenons being raised up from newel-entry at the side of the bottom riser, to newel-entry at the side of the second riser.

Bullseye glass: A small pane of glass containing a centralized, raised (blown) bullseye feature, resembling a slightly-protruding, dimpled bottle-base.
Bullseye window: A small, circular window, either fixed or (traditionally) pivoting.

Bungalow: A one-storey house that seems to have derived its name from either Australia or Hindustan and which originally meant a lightly-built dwelling of temporary construction.

Burr: A sharp metal edge – usually unwanted – in the form of a projecting-lip from the true, square arris of the metal. However, such an edge – very much wanted – is skilfully developed by joiners, on the honed, steel edges of cabinet scrapers.

Burst water-pipes: All water-pipes in exposed, semi-exposed, or unheated positions in a building – especially a roof void – should be well-protected from frost by good thermal-insulation material. It must be realized that if the water freezes in a pipe and turns to ice, it will expand and is extremely likely to burst the pipe – which may not be noticed until the temperature rises and causes the ice to melt.

Butler's sink: See **Belfast/butler's sink**

Butt hinges: These most common hinges, used in slightly different forms, metals and sizes for door-hanging, were (and still are) traditionally referred to as *butts* because their opposing leaves are opposite each other in a closed position, i.e., they butt-up to each other.

Butt-joint: A side-to-side, end-to-end or end-to-side abutment in timber or metal, etc., without any overlapping.

Buttresses: Steeply-pitched – and sometimes tiered – supports that project from the face of external walls, either built of brickwork or stonework in keeping with the built-structure's material. Historically, especially on ancient cathedrals and churches, buttresses were built as ornate features to prevent large, unsupported areas of masonry- or stonework-walls from bowing outwards via their relative slenderness.

    Buttresses might also be added as permanent supports to exterior walls that have developed structural/foundation problems, or to walls that were previously interior party-walls, but have suffered (for whatever reason) the permanent removal of an adjoining building, or buildings. Such removal – like a mid-area dental extraction – can weaken the remaining side-structures.

Buyer's Building Survey: See **Building Survey Report**

BWMA windows: *Figure 17*: These initials stand for British Wood-Machining Manufacturers Association, who – just after WWII – redesigned traditional casement windows and frames to reduce the amount of timber used and to speed up the machining

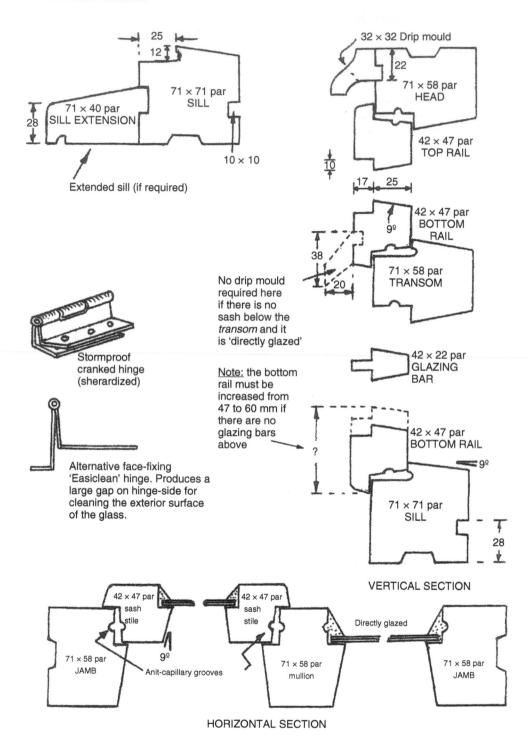

25

12

71 × 71 par
SILL

71 × 40 par
SILL EXTENSION

28

10 × 10

Extended sill (if required)

32 × 32 Drip mould

22

71 × 58 par
HEAD

42 × 47 par
TOP RAIL

10

17   25

42 × 47 par
BOTTOM
RAIL

9º

38

20

71 × 58 par
TRANSOM

No drip mould
required here
if there is no
sash below the
transom and it
is 'directly glazed'

Stormproof
cranked hinge
(sherardized)

Note: the bottom
rail must be
increased from
47 to 60 mm if
there are no
glazing bars
above

42 × 22 par
GLAZING
BAR

?

42 × 47 par
BOTTOM RAIL

9º

71 × 71 par
SILL

28

Alternative face-fixing
'Easiclean' hinge. Produces a
large gap on hinge-side for
cleaning the exterior surface
of the glass.

VERTICAL SECTION

42 × 47 par
sash
stile

42 × 47 par
sash
stile

Directly glazed

71 × 58 par
JAMB

9º

Anit-capillary grooves

71 × 58 par
mullion

71 × 58 par
JAMB

HORIZONTAL SECTION

Figure 17 BWMA casement window details

process. This was achieved by using *comb joints* on the window-frames and the side- and top-hung casement-sashes, instead of traditional mortise-and-tenon joints. This new technique thereby increased the length of the glue-line and the joints' strength. And by rebating the casement-sashes on all four edges (thereby allowing the sashes to protrude) this lessened the timber-sizes of the accommodating frame. Additionally, the comb joints were reinforced by an alloy *star-dowel* driven in from their face-sides. And the seemingly vulnerable, protruding, rebated edges of the hinged-sashes had deep, semi-circular anti-capillary grooves around their inner edges – immediately opposite similar grooves around the opposing surfaces of the side-jambs and the frame's head – to combat ingress of moisture. Sherardized, cranked hinges were used on the sashes and were referred to as *stormproof hinges*.

C

Cabin hook: A metal hooked arm, loosely attached to a fixing plate, with a separate staple fixed in a position to receive the hooked-arm connection to hold open a gate or a door.

Camber: See **Spring/sprung**

Cambered brick-arch: See **Straight brick-arch**

Came/cames (*pronounced cam/cams*):   See **Leaded-light windows**

Capillary attraction: This term refers to the phenomenal manifestation of liquids being attracted to (soaked *up* into, against gravity) porous cellular-material, such as blotting paper (experimentally), or (in reality) most building materials (with an inconspicuous cellular structure), such as bricks, concrete, plaster, timber, etc.

Capillary grooves: See *Figure 3* on page 6: These small, semi-circular grooves (often seen to be opposite each other for maximum effect) can be seen on the side-edges of traditional- and modern-casement windows and window frames. If being pedantic, they should be called *anti*-capillary grooves. This is because their purpose is to interrupt and break the creeping effect of water-seepage between the two close surfaces, scientifically known as capillary attraction or capillarity.

Capped chimneys: See **Deliquescent soot-ash staining to chimney-breast plaster**

Carbon monoxide detector alarm: This inexpensive alarm is not compulsory in one's home, but, for safety's sake, it is highly recommended. Carbon monoxide is a poisonous gas with no odour, colour or taste, that can escape from faulty fuel-burning appliances such as boilers, cookers, etc., and open- or room-sealed fireplaces which have not been properly fitted or regularly maintained by certificated, qualified installers or service providers. Faulty appliances can cause serious illness and death. Unsafe levels of carbon monoxide can be produced when gas, solid fuel or oil does not burn properly, which can occur when fuel is burned without enough air, or appliances need attention. Alarms should carry a BS Kitemark and meet current safety standards to EN 50291.

Carpentry: This on-site activity is usually referred to in two parts: **1)** First-fixing carpentry – such as roofing, fitting and fixing floor joists and floor-decking, etc. – before plastering or dry-lining of walls and ceilings takes place; and **2)** Second-fixing carpentry – such as fitting staircases, hanging doors, fixing skirting and architraves, etc., after dry-lined or wet-plastering has taken place.

Casement fasteners: These items of so-called *ironmongery* are in two forms nowadays; **1)** As used for fastening side-hung wooden casement windows, they have a drop-lever-handled pivoting fastener (originally scroll-ended) that is screwed to the central, opening-side of the casement – and a mortise-holed receiving plate that is screwed to the window's side-jamb, after a mortise slot has been cut; **2)** Plastic-coated and key-lockable casement fasteners are fitted to uPVC side-opening windows and fanlights – which, when turned through 90^0 to open or close, operate the espagnolette-type, concealed shoot-and-slide bolts.

Casement stays: These items of ironmongery are only still necessary on traditional-style wooden casement windows and fanlights, to control and restrict such opening-outwards' windows from being blown open too far. (Note that this risk does not exist with uPVC casements, because their hinge-mechanisms are self-restricting.)

A multi-holed metal bar (the *stay*) with a scrolled or cranked end, is attached to a multi-pivotal fitting that is screwed to the bottom-rail face of the hinged casement-window. To fasten the casement window in an open or closed position, two metal screw-plates with central, upright pins (resembling inverted T shapes), are screwed to the inner top-surface of the sill (or transom) to suit the position of two of the holes in the casement stay. These pins are best fixed in a slightly offset position to affect a degree of *stay-leverage*, giving a tighter fitting of the bottom, opening-end of the casement window.

Casement windows: Traditional, side- or top-hinged windows (hinged at the bottom, as well, but these would have to open inwards), or fixed (un-opening windows, referred to as *fixed lights*), all in casement frames and of wood, uPVC or metal. (For relatively modern, rebated, wooden casement windows, see under **BWMA windows:** *Figure 17*.)

Categories of stairs: As per the Building Regulations 2010, Approved Document (AD) K1, 2013 Amended Edition, references are made to three categories of stairs, as shown below:

Category 1: Private stair (as per Approved Document (AD) K1) for dwellings; Any rise between 150mm and 220mm can be used with any going between 220mm and 300mm, subject to the design formula of 2R+G (covered here separately). The maximum stair-pitch is 42^0. Note that for means of access for disabled people, reference should be made to *Approved Document M, Section 6*, as amended in 2004. And note that for external tapered steps and stairs that are part of the building, the *going* of each step should be a minimum of 280mm.

Category 2: Utility stair (as per Approved Document (AD) K1) for escape, access for maintenance, or for purposes other than as the usual route for moving between levels on a day-to-day basis. Any rise between 150mm and 190mm can be used with any going between 250mm and 400mm, subject to the design formula of 2R+G (covered here separately). No stair-pitch regulation is stipulated, but note that other criteria in paragraph 1.42 of Ad K1: *Access for maintenance for buildings other than dwellings*, may apply.

Category 3: General access stair (as per Approved Document (AD) K1) in all buildings other than dwellings; note that for school buildings, AD K1's preferred going is 280mm minimum – and a rise of 150mm. Any rise between 150mm and 170mm can be used with any going between 250mm and 400mm, subject to the design formula of 2R+G (covered here separately). No stair-pitch regulation is given – but note that for disabled people, reference should be made to *Approved Document M*, as amended in 2004.

Cats-slide roof: This is a variation on a traditional hip-ended, pitched roof above a two-storey dwelling, whereby the roof on one end of the building carries on down past the eaves – but at a lowered pitch-angle, to form a so-called *cats-slide roof* over a room or rooms at ground-floor level. The lowered pitch gives that portion of the ground-floor roof a sprocketed, bell-shape-flared appearance, similar to the base of a playground slide. But much more of a drop-off point that might suit a cat, but not a person.

Cat walk: This traditional term usually applies to narrow gangways provided for building maintenance, such as the boarded area in a roof-loft, leading from a trap-hatch in a ceiling.

Caulking materials: Caulking in building terms is a reference to *certain materials that hold or fix things together in a desired position.* However, nowadays, it also refers to caulking guns that use mastic sealants for sealing the perimeter abutment-joints around window- and door-frames, etc.

Cavity trays: *Figure 18:* If we discount mid-Victorian cavity walls, which were substantially built with a one-brick-thick (225mm/9 in.) inner skin and a half-brick-thick (112mm/4½ in.) outer skin, held together with a criss-cross arrangement of metal wall-ties across a minimal 50mm/2 in. cavity, it can be said that cavity walls – almost in their present-day form (of two half-brick-thick skins) – have been in existence for at least 80 years now. Their popularity, of course, is still linked to the concept of keeping

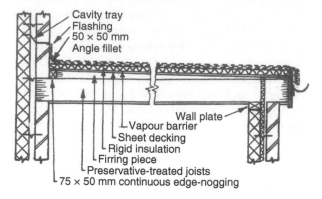

Figure 18 Use of a cavity tray when adjacent flat roof butts up against a cavity wall

a potentially-wet, outer wall away from a dry inner wall, via a cavity, bridged only by essential metal wall-ties.

However, what was overlooked for many years was that when structural abutments (in the form of extensions or outbuildings with flat-, lean-to-, or pitched-roofs) were built against the outer-skin of a cavity wall, such an abutment was causing an acceptably-wet outer wall to become a partly *unacceptably*-wet *inner* wall, via *descending damp*. Hence the origination of upturned, reflex-shaped cavity trays, which were introduced to form a sloping tray across the cavity to act as a DPC (damp-proof course) in the outer-leaf, seated above the turned-in edges of the abutting roof's flashing material. If done success-fully, the now-enclosed wet, outer-wall, would remain a double-dry, double-skin wall.

Cavity-wall-fill extraction: Having always been a believer in a separating space (a cavity) between the potentially-damp outer leaf and the normally-dry inner leaf of a cavity wall, I was not too surprised some years ago when lateral damp problems to the interior surfaces of cavity walls started being discovered. If there is no guarantee for the cavity-infill work, it would be wise to seek the advice of an experienced, independent surveyor, before getting quotes from professional cavity-wall-insulation extractors. The process of extracting involves removing a certain number of single stretcher-bricks in the face-brickwork, just above DPC level and – usually – just above first-floor level, to enable mechanical extraction to take place. The bricks, of course (or new replacements) will be reinstated on completion, but (aesthetically) the property will now exhibit the scars of having suffered damp problems, unless the brick-removal and replacement technique is well-above average in managing to remove the bricks (via removing their mortar joints) carefully enough for reinstatement.

Cavity-wall insulation: See **Cavity-wall problems** and/or **Cavity-wall ties**

Cavity-wall problems: Historically, *double-leaf* (*skin*) walls were originated to be as free as possible from each other, for the purpose of preventing rainwater (that may penetrate through the outer-walls) from bridging across the central-cavity and causing damp problems to the inner-skin walls. Even the criss-cross arrangement of original metal wall-ties was designed with mid-length anti-capillary drip-grooves or – on a later type – butterfly-shaped, wire wall-ties with a twisted, mid-length drip-facility. But, as mentioned under the **Cavity walls** heading, filling cavity walls with insulation took hold in the early 1960s, in an initiative towards energy-efficiency. And, to be fair, there only seems to be a relatively small number of cases where problems with lateral-damp penetration have occurred.

The three most commonly used fillings seem to be **1)** Mineral wool/Rockwool, in the form of blown mineral fibre; **2)** Polystyrene beads or granules – which are bonded together with an adhesive to prevent them from escaping during any future, structural alterations; and **(3)** Foam, either polyurethane or urea formaldehyde.

Note that types **1)** and **2)** are believed to be covered by BBA (the British Board of Agrément) and LABCD/P (Local Authority Building Control Departments/Partnerships).

Cavity wall reveal-closers: *Figure 19*:   This term refers to the essential perimeter-*sealing and damp-proofing* arrangement required around any openings – such as doorways and windows – built into cavity walls. This is because the closed ends of these openings (the *reveals*) bridge the cavity and create vulnerable portions of solid-wall construction. Traditionally, cavity-wall reveals were closed with damp-proof-course material sandwiched between the outer skin and the block-bonded inner quoin – but, nowadays, a variety of pre-formed plastic, cavity reveal-closers, lined with foam insulation (to inhibit cold-bridging), are obtainable and easy to install.

Cavity walls: See *Figure 18* on page 46: Such walls are comprised of an *outer* wall and an *inner* wall with a 50mm- to 75mm-wide, uninterrupted cavity between them to prevent lateral, penetrating rainwater from reaching the inner walls. But, since the early 1960s, when energy-efficiency became a concern, the practice of filling the cavities with a variety of insulation materials has caused incidences of dampness bridging across

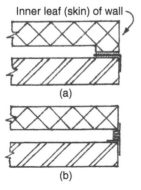

Figure 19 Horizontal sections through (a) traditional, DPC cavity-wall reveal closer; (b) modern, plastic-formed cavity closer bonded to rigid XPS insulation

them and affecting the inner walls. In trade language, these outer and inner walls are referred to as *skins* or *leaves*, hence the term *double-skin* or *single-skin walls*.

Importantly, double-skin walls rely on their structural strength from an arrangement of metal wall-ties that must be built-in across the cavities as the two leaves are being built. The metal ties (which, nowadays – since about 1979 – are non-corrosive), should be bedded into the mortar-joints at every 900mm horizontally on every sixth successive brick-course as the work rises, i.e., at 450mm vertically, but bedded between the ties laid previously below, thereby achieving a structurally-essential criss-cross pattern.

As per the Building Regulations, Part A1(Structure), wall ties should also be provided at the vertical edges of all openings (such as doors and windows), expansion joints and roof verges – spaced at not more than 300mm apart vertically and within a distance of 225mm from all vertical edges.

Cavity-wall ties: *Figure 20*:   These important components, that are positioned to form a criss-cross pattern of support between the two walls, must be so designed to inhibit moisture bridging across them from the exterior via capillary attraction. In the late-19th century, this was achieved by using galvanized iron wall ties with split, V-shaped ends and a twisted, mid-area spiral-shape to act against capillarity. Then, in the mid-20th century, these were superseded by galvanized-wire wall ties, which were twisted in their mid-area, but shaped like butterfly wings at each end.

Both of the above types of wall ties have caused – over a period of time – significant structural damage to cavity walls via a breakdown of the galvanized coating, causing wall-tie corrosion and rust-expansion. The continuous build-up of rust (iron oxide) initially causes identifiable, fine cracks in the horizontal, tie-containing mortar joints. Over time, these cracks become larger and larger and, if left indefinitely, can/will cause partial collapse to the outer leaf of brickwork. Confirmation of the bed-joint cracks being related to wall-tie corrosion is simply gained by checking that the cracks conform to being 450mm/18 in. apart vertically.

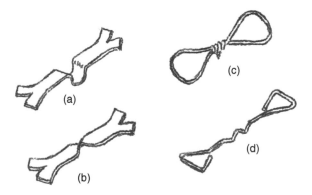

Figure 20 Cavity-wall ties (a) + (b) Late 19th-century galvanized iron wall ties, with mid-area drip facility; (c) Mid-20th-century galvanized-wire wall ties (d) Modern, austenitic stainless-steel wall ties

Currently, modern wall-ties are still made with wire in a butterfly shape, with a double drip feature (be it that the wings are stretched-out and smaller), but – importantly – they are now made with austenitic stainless-steel to BS EN 845-1, and can be used on dwellings up to 10m/32ft 10 in. high. Note, also, that 65mm/2½ in. diameter, plastic *insulation-retaining clips* are being used with these ties, to hold the rigid- or semi-rigid insulation-panels against the inner-leaves of cavity walls, after being friction-clipped onto the wire wall-ties via a V-shaped, centralized slit. This seems to be the sensible way of providing insulation into new cavity walls, without jeopardizing the integrity of the cavity by placing any infill-material against the outer leaves.

C/C: Centre to centre (measurement).

Ceiling joists: See *Figure 21* on page 55: Timber joists which carry the ceiling(s) below them, but which are not usually structurally designed to support habitably-loaded floors above. Therefore, the areas above ceiling joists, are usually limited to being roof voids and the like.

Cellar: See **Basement rooms and/or cellars**

Cellar rot: This **brown rot** wood-destroying fungi, *Coniophora puteana* (originally known as *Coniophora cerebella*), is a wet rot that can develop in humid, damp, unventilated areas. The signs of cellar rot, apart from the brown-coloured timber destruction, can be seen as a matted zigzag, cobweb-like pattern of fine, black strands on the surface of the affected timbers and the nearby brick- or plaster-faced walls. And, if the walls are plastered, these strands may alternatively be hidden on the face of the walls, *under* the plaster. Technically, these strands are the *surface mycelium (hyphae)* and are like vegetable roots, seeking moisture.
Cement: Grey-coloured powder produced from calcined limestone and clay, used universally as a binding agent in sand-and-cement mortar and in concrete, etc.

Cement fillets: See **Sand-and-cement fillets**

Centimetre/cm: One hundredth of a metre, i.e., 1000mm ÷ 100 = ten millimetres/ 10mm/1cm. Note that although centimetres and metres seem to be the main metric division learnt by schoolchildren, the engineering and construction industries (at least) work in millimetres and metres.

Central heating: This term means that the heat being used in an entire dwelling or building is from a central source – a boiler, or heater – which is either one that heats water (a hot-water heating system) or one that heats air (a warm-air heating system).

Cesspits: These are outmoded, underground storage chambers that receive an individual dwelling's waste water and sewage (for a limited period of time), in the absence of a connection to the public sewerage-system. Nowadays, cesspits have mostly been superseded by more sophisticated (and potentially more hygienic) septic tanks. Depending on the number of occupants in a dwelling that use a cesspit, it will need to be emptied by a specialist, sewage-suction company about every six months.

Although traditional cesspits may vary in their usually-rectangular size, there will be a cast-iron inspection cover at ground level – and this may have a grease-seal. The cover will be either set into (or bedded onto) a concrete top, or it may be bedded on the brick walls of the cesspit, which are usually of 225mm/9 in. thickness. And these will have been built onto a concrete foundation/base of (hopefully) 150mm/6 in. thickness. Finally, it is likely that the inner wall-surfaces are sand-and-cement rendered.

Chalet bungalow: A hybrid house, that has a single-storey construction like a bungalow (i.e., with the advantage of de-leafing the leaves and moss droppings from the gutters, via a 3m/10 ft. ladder), but with rooms in the roof void, like a house. The latter being a disadvantage to those who want to avoid climbing stairs.

Chamfered and **Stopped chamfers** on **bricks:**   See **Chamfers/arrises** and/or **Chamfer stops/Stopped chamfers**

Chamfers/arrises: The part-removal of 90° sharp-edges (arrises) on timber, steel or stone, etc., that produces a 45° splay of significant or insignificant size. The former, chamfers, are used for decorative purposes, but the latter, arrises, are commonly referred to in carpentry, joinery and furniture-making, etc., for the purpose of removing a sharp, potentially dangerous edge, that can cut through skin on accidental impact – such as bumping into a door-edge – or can cut/wear through a paint film. For the latter reason, good painters/decorators traditionally rubbed off sharp arrises with glasspaper prior to painting (if the door-hanger had failed to remove them with a smoothing plane).

Chamfer stops/Stopped chamfers: are so-called when they do not remove the whole length of a chamfered arris on balusters and newel posts, etc., and a portion of the edge-material at the top and the bottom is left *square*. This transition from chamfer to square results in a 45° arrowhead shape, if done by hand, or a concaved arrowhead, if done by a router.

Channel: In underground drainage systems, this term usually refers to a 100mm/4 in. wide, semi-circular-shaped open-channel drainage pipe used in traditional manholes (now referred to as *inspection chambers*). Such channels can be of SGW (salt-glazed earthenware) or (according to the chamber's age) of polyvinyl plastic.

Chase: This is a reference to a rough channel or groove cut into the face of a wall, or solid floor-surface, etc., usually for the purpose of concealing a pipe or sheathed cable – after the chase has been 'made good'.

Chimney breast: The three chimney-walls that protrude into a room and contain the fireplace opening and the concealed 225mm x 225mm/9 in. x 9 in. flue(s).

Chimney-breast removal: See **Flying chimney-breasts**

Chimney cap: A clay cap that fits snugly into a clay chimney-pot, to improve or limit the down-draught, and/or to stop birds from nesting in the pot. Chimney cowls of aluminium, stainless steel and galvanized steel, with manufacturer's stated uses, can also be obtained.

Chimney-stack removal: If the uppermost, exterior portion of a disused chimney stack is to be removed and not replaced – it should be capped off with a coping slab, either above the roof, or in the roof-void below. And all fireplace openings that were served by the chimney's flues, must be structurally covered and provided with an *open*-louvred vent – not an optional-closing vent.

Chimney stacks: Prior to the changes in the Building Regulations in 1965, all chimney flues were formed within the brickwork of the chimney stack as it was being built. This meant that the *inner linings* to the 225 x 225mm/9 x 9 in. flues (which were comprised of rendered bricklaying mortar (technically referred to as *parging* or *pargetting* and originally mixed with cow dung) had to be applied very awkwardly at arm's length by trowel, as the stack was being built. And although these chimney stacks and their primitive flues have served quite well over many decades, the flues were superseded by precast, built-in flue-linings soon after the Regulations' change mentioned above. Note, however, that chimney stacks, regardless of old or modern flue-linings, are still a high potential source of descending damp.

Chimneys that lean: See **Leaning chimney-stack phenomena**

Chipboard: This resin-bonded sheet-material (comprised of compressed woodchips) has a popular sheet-size of 2.4m x 1.2m/8 ft x 4 ft and popular thicknesses of 18mm/ ¾ in. and 22mm/7/8 in. It seems to have limited use in building construction, although it is still widely used in its flooring-quality T&G (tongue and grooved) flooring panels, acting as subfloors. In this form, there is a choice of pale-green coloured flooring panels that indicate they are moisture-resistant.

Chord(s):   A technical/structural reference to a trussed-rafter's rafters (top chords) and a trussed-rafter's ceiling-joist ties (bottom chords).

Cill: An alternative British spelling for *sill*, as in *window cill*, although *sill* is more common.

CIOB: The Chartered Institute of Building is a professional body for Chartered Builders and Surveyors and allied construction companies.

Circle-on-circle: This geometrical description was traditionally used to describe double-curvature work, curved in plan-view *and* in elevational-view, such as, for example, a segmental-headed window, in a segmental-shaped bay. Note that such windows required expensive, specially-made circle-on-circle shaped glass.

Circular stair: See **Spiral stair**

C/L: Centre line.

Cladding: **1)** This term literally refers to the outer *clothing* of a built structure, which might be to do with the roof or the walls, but more specifically, nowadays, it is used to refer to the vertical surfaces, i.e., the walls. Traditionally, walls were occasionally partly-clad with hung-tiles, hung-slates, hung-shingles or horizontal (timber or uPVC) weatherboarding. However, on a large number of recently-built dwellings, the upper façades are being completely clad with horizontal, fibrous-cement weatherboarding. But this is probably being done because cladding is less expensive than face-brick façades, or, because such cladding suits the substrate position of the upgraded, thicker wall-panels of insulation.
    2) The term cladding is also emotively connected nowadays to the Grenfell Tower inferno-tragedy in June, 2017. But that was a different form of cladding, designed to provide thermal insulation. The controversial material-mix used for that was technically known as ACM PE cladding (Aluminium Composite Material, with an unmodified polyethylene core).

Cleaning up: This traditional woodworking-joinery term refers to the final removal of any blemishes and uneven joint-abutments of a framed-up and glued item. This is done by planing and/or sanding, after the glued joints are set.

Cleats: Short boards or battens, usually fixed across the grain of other boards to give laminated support to the join. Commonly used in timber formwork (shuttering) and often clench-nailed.

Clench-nailed/clenching: Two pieces of timber, in a cross-grained position, fixed together by round-head wire nails, with about 6 to 10 mm of projecting point bent over and flattened on the timber, either in the direction of the grain for a visual finish, or across the grain for extra strength. Traditionally, such nailing (predominantly used on site-shuttering (formwork) and used for board-fixings on ledged-and-braced doors) was done with *black-iron cut-clasp nails*. Such fixings produced extremely strong connections – which were not always aesthetically acceptable.

Clerk of Works: Traditionally, such positions in the building industry were held by very experienced and technically-qualified people with a general, working knowledge of all trades and good site practices. Usually, such people were ex-apprenticeship-trained tradesmen with an advanced-level of technical-college training and knowledge, who had progressed from trade-foremanship to general-foremanship. General foremen, employed by the building contractor, were traditionally in charge of the whole on-site building project. Their equivalent status in present-day titles would be Site Agents/Site Managers/Construction Managers – call them what you will. Clerks of Work often emanated from having been experienced General Foremen, but the main, critical difference was that the former was employed by the *client* – not the contractor. They liaised with the General Foreman, but kept a very close eye on the quality of work and working practices, kept records and reported to the architect or the site engineer; who also worked for the client. But, instead of Clerks of Work – who mostly fell victim to cost-cutting practices in the 1980s – we now appear to have NDAs (Non-disclosure Agreements), that some purchasers of dwellings seem to be obliged to sign before buying a property – to bind them from disclosing whatever building defects may materialize after purchase. Clerks of Work virtually disappeared in the 1980s, via various changes in building practices related to cost-effectiveness. However, the Institute of Clerks of Works of Great Britain Incorporated, believed to have been founded in the late-19th century, have now changed their name to the Institute of Clerks of Works and Construction Inspectorate (ICWCI) and appear to have attracted more members in recent years – hopefully in readiness for a more responsible regime of building control in relation to non-altruistic practices.

Clinker blocks and clinker bricks: See **Breeze blocks** and **breeze bricks**

Close-boarded fencing: This term usually refers to vertical, featheredged boarding, nailed to three horizontal arris rails fixed into, or onto, posts set in the ground at a maximum of 3m/9ft. 10in. apart. The end-grain feet of the boards must rest on the edges of gravel boards fixed between the posts, to protect the end-grain of the featheredging and to inhibit sagging of the boards and the arris rails. At least one mid-area stake – with a sectional size of 50mm x 50mm/2 in x 2 in – should be driven into the ground and fixed against the arris-rail-side of the gravel board, to inhibit any future lateral bowing-tendency of the board, that may cause loss of support for the featheredge boarding.

Close-coupled WC cistern: With this vitreous-china WC suite, the water-filled, flushing cistern is connected directly to the WC pan – minus a so-called flush pipe. If interested, see also **Low-level WC cistern** and/or **Water closet (WC)**, written herein, for more detail.

Closed (or housed) strings: This refers to a common staircase with side *string-boards* that each have two parallel, uncut edges. Tapered housings are routered/cut into their face-sides to house the ends of the steps with glued and driven wedges at their rear.

Closer: In brickwork terminology, this refers to a site-cut, *half-width 'header' brick*. Such lengthwise brick-cutting is a skilful operation, usually carried out with a bolster chisel. Closers are needed to create a particular bond-pattern (such as *Flemish* or *English* bond), from each quoin (exterior corner).

Coach bolt: This has a thread, nut and washer at one end and a dome-shaped head and partly-square shank at the other. The square portion of the shank is hammered into the round hole to stop the bolt turning whilst being tightened.

Coach screw: A traditional hybrid fixing, that is half-bolt and half-screw. It is commonly 6mm/¼ in. diameter with a square head and a diminishing screw-thread shank. Pre-drilled holes are essential for these fixings.

Coat: This is the usual reference to a number of layers of various materials, such as *one coat of paint, two coats of plaster, three coats of asphalt*, etc.

Cockspur fastener: A traditional-style casement-window fastener, with a cockspur-shaped handle and a face-fixing, mortised side-plate.

Codes of Practice: This is a publication issued by the British Standards Institute (BSI), which describes what is considered to be good trade practices. Their recommendations are not directly enforceable, but indirectly carry the weight of a government quango. However, their publications are well-researched and highly regarded and are widely implemented by building industry professionals.

Coefficient of building-materials: See **Differential expansion**

Cold bridging: See **Thermal bridging**

Cold-deck flat roofs: *Figure 21*:   The salient feature of such a roof is that the insulation is placed *below* the structural deck, between the joists at ceiling level, but with space

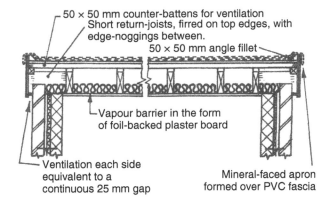
50 × 50 mm counter-battens for ventilation
Short return-joists, firred on top edges, with
edge-noggings between.
50 × 50 mm angle fillet
Vapour barrier in the form
of foil-backed plaster board
Ventilation each side
equivalent to a
continuous 25 mm gap
Mineral-faced apron
formed over PVC fascia

Figure 21 An example of a cold-deck flat roof, when the insulation is between the ceiling-joists

above it to allow for air-circulation. And to avoid cold/thermal bridging, the ceiling insulation should lie across the top of the inner-leaf of the cavity-walls and join up with the cavity-wall's insulation – with care being taken not to block the perimeter air vents. Such vents, positioned in the soffit area, can be continuous or intermittent – and must be on opposite sides of the roof for cross-ventilation. Where this is not possible, then this type of roof should not be used. The openings of the vents, that must incorporate an insect screen mesh, should be equivalent to a continuous 25mm/1 in. gap. Additional vents are recommended for spans over 5m/16 ft. 5 in. and for roofs with an irregular plan-shape. See **Insulation to flat roofs** herein.

Cold legs of water: See **Combination boiler**

Cold-water storage tank (cistern):   Historically, one of these tanks were (and still are in many houses) located in the roof-void (loft area) of a dwelling house, as has been the practice for more than a century, and, in that period, they have been manufactured from a variety of materials, including (in recent decades) asbestos and galvanized steel. According to recent surveys, a number of these two named materials still remain in ill-maintained dwellings.

But, where cold-water storage tanks still exist, they are usually made of plastic, have a fitted plastic lid and are square-, rectangular- or circular-shaped. After installation, they must be well-insulated against frost on their sides and top.

Functionally, this tank holds a supply of cold water for the appliances it serves in an indirect system of cold-water supply. If it feeds the hot water supply as well, it should be large enough to hold at least 230 litres. The tank's ball-valve should be fitted as high as possible and it must be fitted with an overflow warning-pipe of not-less-than 19mm/ ¾ in. diameter.

Cold-water supply: Local Water Authorities (such as Southern Water) are responsible for supplying water to household properties and industries, etc., to a point close to the property, just outside the legal boundary line, where – nowadays – a water meter is usually installed.

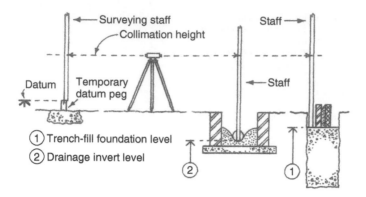

Figure 22 Collimation height

Collar(s):   See *Figure 11* on page 22: In traditional roofing, these are horizontal timber-ties (usually of 100 x 50mm section) that join pairs of common rafters together, theoretically at any point above wall plate (eaves') level – but usually in the mid-rafter area, to act as additional support for the cross-purlins. Note that in loft-conversions/attic rooms, collars are used as (become) ceiling joists.

Collimation height: *Figure 22*:   This term is used in site-levelling with a tripod level and refers to the height of the level's theoretical viewing plane above the original Ordnance datum established above mean sea-level at Newlyn in Cornwall.

Column: An upright, permanent support, designed to carry a structural load, which is usually round, square or rectangular in section and comprised of reinforced concrete, brickwork, steel, stone or timber. Historically, columns were also of square-sectioned and lathe-turned softwood and/or hardwood and can still be found supporting heavily-loaded porches, etc.

Comb- or Finger-joints: These traditional *bridle joints* are nowadays widely used for jointing wooden casement windows (and in recent years, they have been adopted by the furniture-making industry). Unlike the traditional 3-fingered, open-ended bridle joints used in carpentry, comb- or finger-joints have at least 5 layers or fingers, thereby increasing the glue line and the strength of the joint. The odd number of layers achieves the same grain-direction on the face-side and the rear-side of the joint.

Combination boiler: This relatively modern domestic boiler is designed to heat up the water instantly whenever it is required. It also serves a sealed system of hot-water central heating and no separate cold-water storage tank is required. This is a major advantage with this type of boiler, as there has always been some concern about leakages and freezing up of cold-water storage tanks (cisterns) and their pipes housed above one's bedroom areas, in the roof void. However, one disadvantage is the so-called *leg of cold water* that has to be run off (wasted) before the required hot water arrives at a sink, basin or bath, etc.

Commode step: Traditionally, this referred to the bottom step of a staircase that had a slightly curved (convex-shaped) riser- and tread-board, that ended in the outer newel-post face by transcending from the slightly-curved shape to a quadrant-shape.

Common bricks: See **Common brickwork**

Common brickwork: Roughly finished brickwork, built with cheaper, non-facing *common bricks*, to be plastered; roughly pointed (and *bagged-over* with sacking material); or otherwise covered. Such bricks are used as *stretchers* on the inner faces of single-skin, one-brick-thick (225mm/9 in.) walls (as detailed under **Fletton bricks**).

Common-joists: See **Bridging- or common-joists**

Common rafters: See *Figure 11* on page 22: Traditionally, sawn timbers of 100mm x 50mm/4 in x 2 in were used for these load-bearing roof-ribs that pitch up from the wall plates on each side of the roof-span, to rest opposite each other and be fixed to the wall plate at the bottom and against the ridge board at the top.

Compartmentation: This terminology, in the Building Regulations' Approved Document, Part B: Fire safety, refers to the prevention of the spread of fire within a building by designing it to be *compartmentalized* into separate areas, via designated fire-resisting walls and floors, etc. For example, the walls and any floor between a garage and a house, should have a 30-minute fire resistance, and any opening in the wall should be at least 100mm/4 in above the garage floor level and be accessible via a 30-minute FR (fire-resisting) door. And in other buildings, such as a block of flats, compartment walls and/or compartment floors should be constructed between every floor – unless it is within a maisonette.

Approved Document, Part B (amendments), also refers to Fire Safety and Fire Alarm Systems in relation to the design of buildings and speedy evacuation in the event of fire. There is also reference to an upgrading of smoke alarms in *new* dwelling houses, and, in Section B2, there is reference to mandatory *Means of escape* via emergency egress from ground-level *and* upper-floor-level windows (or doors).

Completion certificate (or letter):   When additional building work has been carried out on a property, which was legally done under the jurisdiction of the Local Authority's Building Control Department (or Partnership), there should be a *completion certificate* or letter issued to the property owner, to confirm that the work had been completed satisfactorily. However, builders often neglect to inform the Authority that the work is complete and the property owner is often unaware of the certificate's relevance. Such anomalies usually only become known when the property is being sold and the legal requirement for the completion certificate has been raised by a surveyor or a solicitor.

Completion date: 1) An agreed date given in a building contract for the completion of the specified work. A penalty may be incurred if the builder exceeds this date. However, extra work sanctioned by the architect on W/O's (works' orders), bad weather or other hold-ups – if documented by the builder/site manager – may be taken into account. 2) This term also applies to the legal transaction of buying and/or selling a property, whereby a date is agreed between the seller and the buyer for legal completion to take place, subject to the exchange of contracts and the receipt of the purchase price.

Compo: This is a traditional bricklaying term for *mortar*, being a site-abbreviation of the mortar's *compo*sition.

Composite construction: *Figure 23*: This refers to different materials used in conjunction with each other. One example of this is a relatively modern *Beam-and-block* floor, which is often used nowadays for suspended ground-floors in dwelling houses. The inverted T-shaped beams are of factory-made precast reinforced concrete, which accommodate successive rows of dry-laid, aerated building-blocks on the beams' protruding, lower edges – the blocks, which are flush to the top edges of the beams, are overlaid at a later stage with rigid sheets of insulation, a polythene vapour barrier and a T&G flooring-panelled floating floor. However, although this example of *composite construction* is sensibly separated, it must be understood that varying rates of thermal expansion and contraction exist between different materials – and structural problems of cracking, flaking, bulging, etc., can be caused by material-mixing, i.e., close-attachment of mixed materials.

Composite doors: This adjective refers to these doors being comprised of a *composition* of different materials, built up internally to achieve a solid core, with superior strength and good security. Although these materials may vary via different manufacturers, they

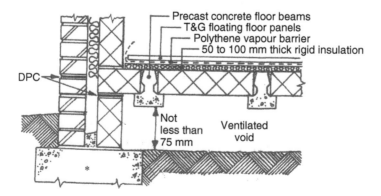

Figure 23 Beam-and-block ground-floor construction is an example of good 'composite' construction. *Note the Trench-fill, strip foundation

include metal-trim, foam (for thermal performance), laminated wood (like plywood, but with much thicker cross-layers), GRP (glass-reinforced plastic), wood-grained poly-ethylene outer-skins, etc.

These doors usually have multi-point locks which are comprised of two hooks, two so-called finger bolts and a deadbolt – all operated by one upwards turn of the door's lever handle and then locked by the turn of the key inserted in the keyhole's lock lever-furniture.

Three types of locking arrangement are usually available: **1)** *Lever/Lever lock*, which is always locked via the key, to avoid locking yourself out; **2)** *Split spindle lock*, whereby you can open the door and exit by using the lever handles, but you will need a key to re-enter the house; and **3)** *Slam lock*, for fast exit. When you close (or slam) the door upon leaving, there is no need to lift the handle *or use a key* to lock up – but you will need a key to regain entry and to lock up from the inside of the house.

These doors are made with a variety of small, obscure-glazed apertures, either with a British Kitemark for toughness/fire resistance, or with a European CWG EN12150 LOW-E (emission's mark).

Composite slates: As mentioned above, this adjective refers to these items being comprised of a *composition* of different materials, manufactured to produce artificial slates that resemble natural, quarried slate. Nowadays, there are also composite *tiles*, made of a mix-ture of materials other than traditional clay or concrete. Whether these products will last as long as natural slate or clay tiles is unknown. But their known cost-effective advantages are: **1)** they are lighter in weight, making them easier and quicker to lay; **2)** their sizes are more uniform and they can be cut easily and cleanly; **3)** they are cheaper to produce than slates from depleted quarries. Composite slates have been in existence for many decades, but they originally contained chrysotile fibres (also known as *white asbestos*). This is now banned from use in the UK, so modern composite slates (such as Marley Eternit fibrous slates) and tiles are comprised of a variety of other materials, including synthetic resins, GRP (glass-reinforced plastic) fibres, cement, concrete and clay – such as the Weinerberger's Sandtoft tiles and the interlocking Rivius slates made from clay.

Compression area: See **Tension and compression areas**

Concave: Shaped like the inner surface of a *sphere* (or *cave*).

Concealed hinges: These are commonly fitted on kitchen-unit doors, etc., nowadays, but the type referred to here are secreted *edge-fitting hinges*, made of die-cast zinc, which are suitable for internal, residential and commercial room-doors. I have seen them used on a secret door that led to a concealed kitchen in a wall-panelled boardroom; the shallow, imitation shelves on the face of the door carried equally-shallow and realistic-looking hardback books.

Concentric: Sharing the same centre point.

Concrete: An easily-produced, stone-like building material which consists of mixing various ratios of sharp sand, ballast (gravel), cement and clean water. On large building contracts, these important ratios are usually strictly controlled by mechanical batching – and testing – or by transported, ready-mixed and pumped concrete from specialist suppliers.

Concrete cancer: This phrase has been coined in recent years to refer to a deterioration of reinforced-concrete components, whereby surface-damage can appear, initially in the form of hairline cracks. Over time, the cracks enlarge and the concrete surface begins to break up (spall) and fall away in small or large pieces. In all known cases, this seems to be related to the pressures created by corrosion and rust-expansion build-up of the reinforcing-steel buried within the tensile (bottom) areas of the concrete. Such damage to the reinforcing-steel is now known to be caused by carbonization, caused by atmospheric pollution.

Concrete-dusting: This term refers to concrete-floor surfaces – such as in a garage – that, during normal wear-and-tear of walking and sweeping, are always producing a cementitious dust. This is reckoned to be caused by an excessive amount of water being added in the mix, or by an excessive amount of surface-laitance having been brought up to the surface by over-tamping the newly-laid concrete.

Concrete gutters: See **Finlock-type guttering**

Concrete houses and bungalows: Many such dwellings were built in the UK, via local authorities, to meet a demand for the speedy erection of dwellings in the post-war period of 1950 and (as outlined above under the heading of **concrete cancer**) many of them have suffered significant damage and have since been demolished. Those most affected by this were built with precast, reinforced-concrete panels containing ferrous-steel reinforcing bars and bolted connections.

Other concrete houses and bungalows built in this period, were built *in situ* (i.e., *on site*) with *no-fines concrete* – a non-standard mixture, described below. And, although not free from all criticism, such houses (minus the ferrous-steel) survived better than the precast, reinforced-concrete panelled dwellings outlined above.

Almost uniquely, preformed, full-height wall-shuttering panels (or, *formwork* panels) were used on site and were filled with the no-fines concrete. Such concrete (not in use or well-regarded nowadays) was comprised of a cement-slurry and only one other ingredient: *ballast* (also referred to as gravel), of a sieved-size of not more than 19mm/¾ in diameter, nor less than 9mm/⅜ in diameter. There was (unusually) no fine aggregate or sharp sand, as used in regular concrete. Such an unorthodox mixture for whole-wall construction, when set, produced an aeriated, honeycomb structure, like Aero chocolate, but less dense. However, their exterior surfaces were rendered and their interiors were either dry-lined or lined with traditional two- or three-coat wet plastering.

Their problems are mainly to do with **a)** Condensation due to poor insulation values; **b)** Poor heat-retention; **c)** Thermal cracking of external rendering, causing lateral damp-penetration; **d)** Possible thermal-bridging in many places; **e)** If and when identified by surveyors for a purchaser, concrete dwellings can be difficult to mortgage.

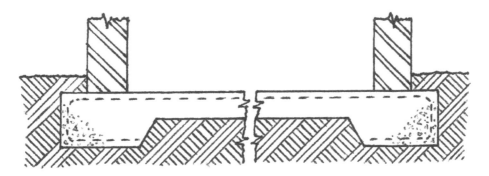

Figure 24 Reinforced concrete raft foundation

Concrete lintels and beams: Prior to the mid-20th century, reinforced-concrete components – especially lintels over doorways and windows – were often cast *in situ* (i.e., *on-site*), via temporary timber-formwork (often referred to as *shuttering*). When such lintels were needed over windows in cavity-wall construction, they were often required to be *in situ boot-lintels* (covered here separately), to enable the toe of the boot-shape to occupy only one rise (course) of face-brickwork. When such lintels and beams were specified, the contractors were given details of the steel-bar reinforcement required, usually in the form of stirrup-linked cages. As mentioned in this book under **Boot-lintels**, such lintels are now likely to be displaying signs of deterioration.

Concrete-raft foundations: *Figure 24*:   It is sometimes structurally possible to use a raft foundation for a relatively simple, single-storey construction, or for areas with weak subsoil. They are comprised of reinforced concrete slabs of 100mm/4 in to 150mm/6 in thickness, with turned-down (integrated), foundation-edges all around, which also require to be reinforced with steel bars or stirrup-shaped steel cages, near their bottom, tensile areas.

Concrete spurs: This refers to precast reinforced-concrete posts of varying lengths, usually with two bolt holes in their upper region, which are used for supporting weakened and/or deteriorated fence posts, via coach bolts, temporary bracing and deposits of stiff concrete at their excavated base.

Concrete tiles: See **Interlocking tiles**

Condensation problems in dwellings: All of the air around us contains water vapour to a lesser or greater degree. As far as condensation is concerned, the amount of water vapour and the temperature is critical. Water vapour contributes to the total pressure of the air; the higher the moisture content, the greater the pressure.

The quantity of water vapour in air is measured in several ways. One method is by *relative humidity* (RH), which is closely linked with temperature. Warm air can hold more water vapour than cold air, but at a certain temperature it can only hold so much. When saturated, its RH (relative humidity) is 100%.

The temperature of the air at which saturation occurs is called the *dew point*. If warm air is cooled, the RH increases until it reaches the dew point. Any further drop in its temperature means that the air is overloaded and the water vapour condenses on colder surfaces (such as walls, glazing and soft, furnishing materials and bedding) where the temperature is at, or below dew point.

Condensation problems can only be overcome if *ventilation, thermal insulation and heating* are fully understood and implemented. A combination of these is essential. For example, thermal insulation without heating does not help much – and heating without ventilation does nothing to help dry the rooms. The margin between condensation occurring or not is often very narrow. Dwellings left unoccupied and unheated during the day are especially vulnerable. Household activities causing heavy concentrations of water vapour are then carried out within a short period during the evening, causing the internal air to become saturated with water vapour which condenses on cooler surfaces.

Inside a dwelling, the water vapour tends to build up. In bedrooms, adults breathe out about 18 grammes of water vapour hourly whilst sleeping – and much more when moving around. The average family is reckoned to breathe, cook and wash about 12 litres of water vapour per day into their home. The biggest source is from daily cooking, washing clothes, bathing or showering. Although these activities are confined to specific rooms, high humidity and vapour pressure are produced and they tend to diffuse into the drier air in other rooms very quickly, if allowed to do so.

On impervious surfaces, such as glazing, the condensation will be obvious as water droplets, but on absorbent surfaces, such as plastered or dry-lined plasterboard walls, the signs are not always as clear. Moisture is still deposited, but instead of remaining on the surface, it passes into the material where fungus can germinate and grow, resulting in a dark-staining attack of mould growth – which can also be found on clothes and bedding.

Ventilation through an open fanlight or an *open* trickle-vent in the frame-head of a casement window is very helpful, but should not be wholly relied upon to cure condensation problems. This is because natural ventilation is not as good as mechanical ventilation in the main vapour-producing areas of the kitchen and bath/shower room. Vapour extraction via mechanical extractor fans in these areas is essential. However, a partly-open fanlight at night, in an occupied bedroom, lessens the risk of condensation and mould growth.

Finally, in cold weather, at least a low degree of background heating should be on all the time. Long periods of low heating are far better than short bursts of high temperature heating.

Condensing boiler: This design of boiler provides up to 20% increased efficiency, when compared to similar traditional types of boilers. This is brought about by the heat-source gases that rise up through the central, primary heat exchanger in the boiler, being diverted to the left and the right at the top, to each go back down the inner-sides of the boiler, over the surface of secondary heat exchangers. This reduces the temperature and causes the water vapour to condense and release the latent heat that would otherwise escape with the flue gases.

Conditioning: This refers to balancing the *moisture content* of timber items such as skirting boards, architraves, flooring, doors and sheet material, etc., to the humidity of the rooms and areas in which they are to be fixed or fitted, by leaving them in or near these areas for at least 48 hours. However, note that the conditioning of hardboard sheets, if to be used as an overlay on wooden-boarded floors, acting as an underlay for fitted carpet, etc., should be wet-sponged thoroughly 24 hours before being nailed stretcher-bond pattern to the floor. Failure to carry out such conditioning, will likely result in excessive bulging of the abutted boards after about 12 to 24 hours.

Condition Report: See **Building Inspections**

Conduit: A metal pipe for housing electrical cables; although nowadays, plastic and fibre tubes seem to be mostly used.

Conservation areas: This usually refers to built-up areas that have been designated (by a local authority's planning department) as being of architectural or historic interest and which are therefore protected from any alterations which might destroy their character. As an example, in such areas, early-period properties with vertically-sliding sash windows housed in traditional wooden boxframes, are not usually allowed to be replaced with new uPVC casement windows.

Conservatory/Orangery: Both of these are *greenhouses*, really; the former for growing and displaying plants, the latter for growing oranges/trees in a cool climate. Mostly, nowadays, conservatories in the UK seem to be built as extensions/usable rooms to a dwelling. Even the glazed roof-slopes are now being boarded over internally, in a seemingly futile attempt to improve their energy efficiency – or to give the illusion of being more like a habitable room.

Consumer unit: Nowadays, this is usually a plastic, wall-fixed box, located near the electric meter, which contains the switched MCB (miniature circuit breaker) fuses that protect all the electrical circuits. It also contains an isolating switch that can be used to cut off the power to the entire building.

Continuous (wreathed) handrailing: *Figure 25:* This refers to a geometrically-*wreathed* – or partly-wreathed – handrail over a geometrical stair, where mid-area newel-post supports are not used (unless they are lathe-turned with projecting spigots to act as dowel-inserts into the underside of integrated handrail newel-caps). Continuous handrailing is also referred to as *geometrical handrailing*.

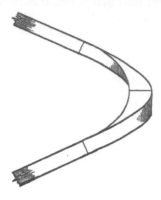

Figure 25 Wreathed handrailing above a Wreathed stair-string

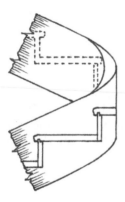

Figure 26 Wreathed stair-string to a geometrical staircase

Continuous (wreathed) string: *Figure 26*: This refers to a geometrically-wreathed – or partly-wreathed – stair-string board of a geometrical stair, where mid-area newel-post supports are not used – and the continuous handrailing is also wreathed, or partly-wreathed.

Contract documents: In contractual Building, these form the legal and binding contract between client and contractor, and consist of the Bill of Quantities, the Specification, the Drawings and the general conditions of the contract.

Contracts Manager: See **Site Agent**

Conventional boiler: A boiler that serves the heating system only.

Converted-houses into flats: See **Flat-conversions**

Converted timber: This is usually a reference to seasoned timber that is re-sawn (converted) to a different sectional size after being purchased.

Convex: Shaped like the outer surface of a sphere.

Coping(s): See *Figure 14* on page 30:   These are usually referred to as coping stones and are used for protecting and/or finishing and weathering the top surfaces of low-rise brick walls. They may be of various materials, such as bricks-on-edge, natural stone, precast, artificial-stone, or precast concrete. The precast copings are usually reinforced with one or two 6 mm/¼ in. reinforcing rods. With the exception of brick copings, which are finished flush to the faces of a wall, the others listed above may have side projections of about 25 mm/1 in., which may be weathered with anti-capillary drip grooves. Also, if the copings are protecting the tops of certain walls – such as parapet walls to a roof – they may be bedded on a DPC (damp-proof course) and have a wedge-shaped, weathered slope to their top surfaces, or two wedge-shaped slopes (one from each side) and be referred to as *saddle-back* copings.

Coping stones: See **Coping(s)**

Copper and plastic pipes below floors: Building- and design-experience should teach us that not enough care and thought are given to inaccessible plumbing services below floors. Hot- and cold-supply pipes do develop faults (usually at their joints, but some-times from corrosion) and start leaking below floors that are usually inaccessible. And, of course, this is exacerbated by modern-day overlay flooring in the form of engineered hardwood, plastic laminates and ceramic floor-tiles, etc.

Traditional square-edged floorboards are not too difficult to remove. But even so, areas of boards (with pipes below) could be screwed, instead of nailed – as was the usual practice with pipe casings above the floors. Such screwing would even indicate where the pipe-runs can be found.

A similar arrangement of selected-board screwing could be done with tongue-and-grooved floorboards, except that the underside lipped-edge of the groove of two adja-cent boards would have to be removed, to enable such boards to be prized up on one side only, (where the lipped edge had been removed) and pivoted out. As this lower edge of the board's groove is (or should be) of less thickness than the one above, removal of it could quickly be done by a traditional *chisel-chopping* technique, as used on hinge-housings – assuming that no power tools were available.

Pipes laid and buried in sand-and-cement floor screeds can also be problematic and cause major problems, especially when such screeded floors are overlaid with traditional pitch-bonded parquet floor-blocks, etc. Leaking pipework in such situations has some-times been discovered to be as a result of metal-corrosion, invariably caused by non-protective wrapping of such pipework. Alarmingly, first signs of leakage can be wrongly diagnosed as rising damp, seen rising up on the internal plastered wall-surfaces, just above the skirting boards. However, if such rising damp is also detected on the inner, partition walls, this is usually a good indication that a major problem exists somewhere within the floor screed. As first stated, more care and thought need to be given regarding inaccessible plumbing arrangements.

Copper slate: See **Lead-** or **copper-slate**

Copper wedges: These are small pieces of copper, folded several times to increase its thickness, and used as driven-in wedges to hold sheet-copper upturns and dressed-copper flashings into raked-out, horizontal masonry joints.

Corbelling: *Figure 27:*   This is a reference to building-out from the face-side of a brick wall in successive, incrementally-projecting, stepped courses. As an example, such corbelling was often used on each side of gable-ended walls, where its purpose was to conceal the ends of the eaves' projections. For this particular use, corbelled *tiles* were also used. However, corbelling was also commonly used to *gather-over* (another brickwork term) the brickwork of two separate chimney breasts on a gable- or party-wall, that were joined together within the roof-void (loft), before they emerged above the roof as a single chimney-stack.

Corbels: These are cantilevered projections from the face of a wall, usually in the form of bricks, hewn-stones, corbelling-irons or stepped brickwork, built-in to support the ends of purlins or beams, etc. Traditionally, corbelling-irons (with a 50mm x 6mm/2 in. x 0¼ in. section), were built-in or driven into a brick-wall's horizontal bed-joint to support the underside-edge of ridge boards fixed to the walls of mono-pitched, lean-to roofs. Corbelling-irons/plates, often with hooked ends, were really the forerunner of modern joist hangers.

Cornice blocks: See *Figure 5* on page 9: Doorway-openings in many *grand houses* and public buildings of yesteryear were often embellished with fluted or reeded architraves

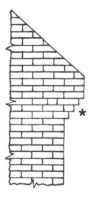

Figure 27 *Brick corbelling to a gable-end wall

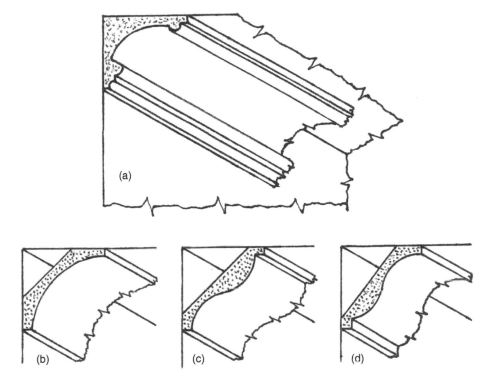

Figure 28 (a) Traditional cornicing; (b), (c), and (d) = Modern coving; (b) Cavetto-shape; (c) Cyma-reversa; (d) Cyma-recta

around their top and sides, that were attached at their base to ornate *plinth blocks* – and at their heads, instead of common mitred intersections, they were butted up against so-called *cornice blocks*. The blocks were square-shaped in elevation and slightly wider and thicker than the architraves. Their faces were often carved or routered out to a decorative pattern.

Cornicing and/or **Coving:**   *Figure 28*: Cornicing is the traditional term for ornate, solid-plaster mouldings that were tediously formed *in situ* with wet plastering techniques and hand-made *running moulds* at the junctions between walls and ceilings. But nowadays, labour-intensive ornate cornices have been superseded by factory-made, precast plasterboard **coving**, which is adhered/stuck into the junctions between walls and ceilings, after being mitred into the corners and at the quoins of a room. Plasterboard coving is either a concaved *cavetto*-shape, or of a more ornate shape known geometrically as *cyma reversa*, which has a **convex** shape that runs along its top face, joined by a **concave** shape that runs along its bottom face. However, the cyma reversa shape can be changed to a *cyma recta* shape, if desired aesthetically, by simply turning the moulding over by 90°, so that the concaved shape is at the top, instead of at the bottom.

Apart from having a decorative feature, a cornice or coving's main, original purpose is/was to conceal future cracks that may – and often do – manifest themselves at the ceiling-and-wall's junction, due to so-called **differential expansion** between the two structures.

Corrugated-metal fasteners: These are shallow-depth joint-fasteners, used to hold glued frame-joints together, such as mitred, right-angled butt joints of simple, ply- or hardboard-clad cupboard doors. They were also referred to as *wriggle pins*. Two are driven in across the face of the joint on one side, then – essentially – one in-between these needs to be driven in from the other side.

Cottages: Small, simply-built houses, usually found in rural areas and seemingly built originally for a local farmer's staff members, labourers, or other tied workers of local businesses, etc. Nowadays, they are usually privately owned by the occupier.

Counter-battens: *Figure 29*:   This term usually refers to timber battens (of a 50mm/ 2 in. x 25mm/1 in. section) that are used on pitched-roof surfaces that have been clad with **sarking**. This Scottish term refers to closely-fitted square-edged boards fixed over the whole raftered roof, to create a more solid structure for the roof-cladding. Sarking is thought to be mostly used in Scotland, but it can occasionally be found to have been used in other regions of the UK. However, apart from stabilizing a roof, sarking also adds a good degree of thermal insulation. Note that such roofs built prior to the early 1930s – unless they have been retiled at a later date – will not be overlaid with a waterproof fabric.

After the roofs have been clad with sarking (which would be laid like square-edged floor boards, with staggered end-joints – but with little regard for neatly-close side-joints), the surfaces should be covered with a waterproof fabric and the *counter-battens* would be fixed at right-angles to the eaves, on top of every under-lying rafter. This counter-battening technique allows the tiling- or slating-battens to be fixed across them at their gauged spacings, thereby providing open areas of roofing-underlay below them, to allow any ingress of rainwater to run down the slopes to the eaves' guttering.

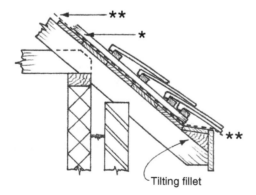

Tilting fillet

Figure 29 *Counter-battens on a sarking(boarded) roof, above ** a water-resistant roofing fabric

Counter-flap hinges: As their name suggests, these hinges were and still are used for hinging counter flaps in commercial outlets, such as public houses and shops, etc. Their leaves are a butterfly (double-dovetail) shape in their opened-up flat position and they are produced in polished brass or stainless steel, etc. They are housed into the top of the counter (or bar) and the opening flap, to achieve a neat, flush finish.

Course: One rise of mortar-bedded bricks, blocks or tiles, etc., laid in a row or a course.

Cover fillet: A square or moulded-edge timber strip, used to cover second-fixing joints, etc. Note, also, that plastic cover-fillets are also used nowadays, to be adhered around the inner edges of uPVC replacement windows and doors, etc.

Cover flashing (to interlocking tiles): *Figure 30*:   This consists of a right-angle shaped length of lead (or other metal) flashing, that is dressed over the contoured surface of the tiles and covers and protects the tile-edges that abut chimneys or parapet walls, etc. Importantly, the top of this one-piece flashing is marked out and cut into saw-tooth shaped steps (with top-turned edges), that are *dressed-and-wedged* into raked-out bed-joints in the brickwork, prior to being sand-and-cement pointed.

Coving: See **Cornicing** and/or **Coving**

Cowl: This usually refers to a circular louvred-cover, of metal or clay, fitted on a chimney to regulate the draught or to offset descending dampness. Cowls also inhibit birds from nesting or becoming trapped in the clay chimney pots.

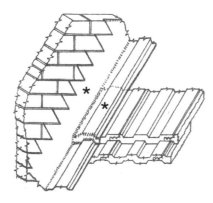

Figure 30 *Lead cover-flashing to tile-abutment of interlocking concrete-titles against a brick wall. Note that on slate or plain-tiled roofs, lead 'soakers' and separate, stepped apron-flashings are used

CPD: 1) Continuous **P**rofessional **D**evelopment. Such periodic updating of knowledge being a requirement of all professional bodies; **2) Cpd** Cupboard.

Cracked-wall monitoring via Crack Monitors: This form of monitoring (also referred to as 'tell-tale monitoring') is usually carried out by surveyors or structural engineers – but can also be done by builders. Simple, durable and vandal-resistant polycarbonate Crack Monitors can be used for monitoring structural cracks in concrete floors or masonry walls. There are also relatively simple and inexpensive 18mm/¾ in. diameter stainless steel Crack Monitoring Discs available for fixing with Araldite on each side of a crack. These tell-tale monitors are pierced with small, central holes for easy measuring between pairs of crack-straddled Crack-Discs, via an inexpensive vernier calliper gauge, that can be purchased with the Monitors, as per a www.YorkSurvey.co.uk catalogue, or other such outlets.

Cracked walls: See **Thermal movement**

Cracks in concrete lintels or beams: Continuous or intermittent cracks that run along the length of reinforced lintels or beams, usually found in the lower, tensile area, are extremely likely to be caused by ferrous-metal corrosion and rust-expansion of the reinforcing rods. In severe cases, the rust-expansion can cause pieces of concrete to break off – which is referred to as *spall* or *spalling* – and undermine the lintel or beam's structural integrity. See **Concrete cancer**, covered above.

Cracks in face-brickwork above windows: In recent decades, a great number of soft-wood casement windows of traditional and BWMA/stormproof-design, with single glazing, have been replaced with uPVC casements and sealed-unit double-glazing. This is understandable from an energy-efficiency and low maintenance point-of-view, but this upgrading/removal of pre- and post-WWII windows can inadvertently cause horizontal mortar-joint cracks in the outer-skins of cavity-wall brickwork, just above the uPVC window-heads.

Occasionally, these long bed-joint cracks were found to be linked to a few stepped joint-cracks at their extremities and, after close inspection, it was confirmed that there were no lintels above the windows to take the brickwork load which had previously been supported by the softwood window-heads. This apparent unawareness of a lintel-deficiency by the uPVC-window fitter (and a number of surveyors, to my knowledge) may be due to the fact that present-day Catnic-type steel-lintels (with thin front-flanges that carry the outer, brick-skins) are barely discernible, so were assumed to be present. Or, may be due to the window-fitter's or surveyor's unawareness of the historical fact that it was a common, accepted practice in the building-industry of yesteryear to build the cavity-wall's outer-skin of brickwork directly onto the softwood window-heads. This was because they were regarded as *wooden lintels*. Also, the projecting *horns* (head-and sill-extensions) of about 75mm/3 in. length, were left on by some bricklayers and built-in to the brick-reveals, thus providing – at the head of the frame – the required bearings to support the frame's head, acting as a wooden lintel. Also, the head-horns and sill-horns were built-in to secure the frames.

Note that the horns on frames were initially left on by the joiners to protect the frames during transit to the site. And, to my certain knowledge, some bricklayers sawed them off and others left them on. Either way, the heads of these yesteryear casements were used to carry the stretcher-bond brick-courses, or brick 'soldier' arches.

Therefore, when appraising mature properties that display uPVC windows, a clue to the prior removal of softwood casements that may have been built-in with projecting horns (and the likelihood of lintel-deficiency), can usually be detected at each end of the uPVC sill, where the removal of the old softwood sill-horns (previously built-in to the brick-reveals) will have been 'made-good' with *large lumps* of sand-and-cement mortar – or, on better-class work, with small offcuts of brick bedded-in neatly with mortar.

Prior to modern-day galvanized-steel lintels that support both the inner- and outer-skins of cavity walls – and prior to a variety of reinforced-concrete lintels, of which, plank lintels still remain – the inner-skins of blockwork were built onto softwood lintels with a sectional size of at least 100mm x 65mm/4 in. x 2½ in.). So, historically, wooden lintels did (and, in old buildings, still do) exist (although oak was used for external lintels).

Therefore, using softwood casement-frame heads as lintels was an extension of a historic practice. And, from a weight-loading point-of-view, these frame-heads usually had/have mid-area support in the form of one or two mullions, acting as props. Another structural factor to be borne in mind is that the actual load above windows- or door-openings is not excessive. It is limited by the stretcher-bond brickwork's bonding pattern between the bearing points, which takes the form of theoretical isosceles triangles with approx. 34° stepped (corbelled) sides – and the remaining, central loads are *usually* self-supporting, via their brick-bonding and mortar-adhesion. Of course, this does not excuse the omission of a bonified lintel.

Finally, to summarize, mortar-bed-joint cracks in brickwork just above replacement uPVC windows could be related to the non-existence of a lintel and the existence of a fitting-tolerance gap above the uPVC window's head – which will have robbed the suspended area of brickwork of its previous window-head support.

Cracks in face-brickwork above windows and doors: See **Soldier arches**

Crack-Stitching: (Traditional method): Cracks in exterior brick and/or stone walls of dwellings and other buildings are caused by a number of issues, including foundation problems, bulging walls, metal wall-tie corrosion and rust-expansion (in cavity walls) – or a similar form of cracking to brick soldier-arches over windows and doorways, caused by corrosion and rust-expansion of early-20th-century reinforcing-bars and ferrous links between the arch-soldiers – and all are covered herein separately, under different headings. But, the majority of cracks that develop in dwellings and buildings (internally and exter-nally) are well known technologically to have been caused by a more common and ubi-quitous phenomena of *natural thermal expansion and contraction of materials*. Hence the relatively-new practice (in recent decades) of forming full-height, vertical, mastic-filled, narrow expansion joints within the length of overlong structural brickwork façades.

From an appraisal point-of-view, one must learn to distinguish between cracks caused by subsidence, or other faults and the more common cracks caused by thermal movement. With regards to the latter, it helps to realize that when thermal expansion has taken place and contraction follows (say, to a front elevation comprised of upper- and lower-windows and an entrance door, within a façade of stretcher-bond brickwork), the structural movement is likely to take routes that offer the least lines of resistance, so causing damage to the brickwork between the apertures. Hence, the diagonal cracks commonly seen between the corner of a window-sill of the storey above and the nearest corner of an offset window- or doorway-opening of the storey below.

Note that assessment of different degrees of damage in BRE's Digest 251 mostly recommends that cracks up to 5mm wide should be repointed – but beyond this, cracks between 5mm and 15mm are recommended to be opened up and *stitch-bonded*. This involves opening up the brick-joints on each side of the crack to enable the damaged bricks to be removed, but also to create an indented bonding pattern across the crack. Then, (if at all possible) matching bricks should be selected, bedded and stitch-bonded across the crack.

(**Modern methods**): As explained in the traditional method of crack-stitching, cracks in exterior brick- and/or stone-walls of dwellings and all other similarly-built buildings, are caused by a number of issues – which are covered here separately in this book under their relevant headings. But historic treatment of structural crack-damage seems to have been either superficially cosmetic or very structurally involved and therefore expensive.

However, seemingly reliable and cost-effective means of repairing and stabilizing cracked walls are now available from companies such as HELIFIX at www.helifix.co.uk. Such products are available direct from Helifix or via outlets such as Screwfix, who give the Helifix Crack-Stitching kit a four-star trade rating. Their kits for crack-stitching repairs essentially contain two components: **1)** spiral-shaped stainless-steel Helibars of 6mm or 4.5mm diameter (the latter for bedding into cementitious grout in *thin* mortar-joints), the former with a 10 kN tensile strength, are for bedding into grout in 10mm mortar joints. These helical-shaped Helibars are in lengths of 1m, 1.5m and 2m., in packs of 10 for cutting to length on site; **2)** Helibond cementitious grout, which must be used for embedding the Helibars into the raked-out mortar joints – is obtainable in 3 litre plastic tubs, which is said to cover 10 linear metres of crack stitching. Incidental tools required are: **1)** a Helifix Pointing Gun CS with a mortar nozzle and **2)** a HeliBar Insertion Tool.

Other applications available from Helifix include: Repair of bay windows; brick arches; cracks near corners and openings; stabilizing bowed walls; reconnecting cross-wall T-junctions; overcoming foundation-settlement – and replacing cavity-wall ties with Helifix DryFix wall ties, Cem Ties, RetroTies, or ResiTies.

For full Product Information, Case Studies and downloadable Repair Details, go to: www.helifix.co.uk/products/remedial-products/crack-stitching.

Cradle: This refers to a boat- or trough-like structure, suspended on hand-controlled rope- or cables for vertical and/or lateral movement on the exterior side of a building. Usually for the purpose of building maintenance and/or repainting operations.

Cradling: Timber grounds formed around steel joists (RSJs), beams, etc., to act as a fixing medium for plasterboard, etc., linings.

Cranked hinges: See *Figure 17* on page 42: This description refers to two types of cranked (L-shaped), sherardized hinges, that were produced to fit the protruding faces of so-called stormproof casement-windows. One of these L-shaped hinge types was mostly concealed within the window's rebated edges, but the other, which was of an inverted T-shape with stepped fixing-plates, was an alternative face-fixing 'Easiclean' hinge, which produced a large gap on the hinge-side for cleaning the exterior glass surface.

Creep: This term refers to sheet lead used on flat roofs, or bay roofs, etc., which, due to thermal movement in relation to the lead-thickness, has the potential to *creep* (move laterally) over a period of time. This phenomenon – which can cause areas of the substrate to become exposed – is only usually discovered when descending damp is discovered in the habitable room(s) below.

Crenellation: A reference to a regularly spaced and repeated indentation, such as at the top of a parapet wall, or as a crenellated soldier-course feature to a garden wall – such being a reference to a battlement wall – which gave partial concealment.

Cripple rafters: *Figure 31*: These are short rafters with a plumb-cut at the head (to fit against a ridge board) and a double (compound) splay-cut at the bottom (to fit – in diminishing pairs – against a valley rafter). They gained their crude (non-PC) name historically by being cut off at the foot (i.e., minus the usual birdsmouth feature and eaves' projection).

Critical-path-analysis chart: See **Progress charts**

Crittall casement-windows and doors: The name Crittall in the window industry is as synonymous with steel windows in the UK as the name Hoover is to vacuum cleaners. And yet I came across a company with this original name that produces various styles

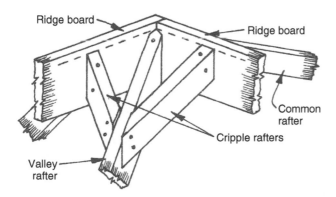

Figure 31 Traditional roofing members

of upgraded steel-windows and doors, with double weather-seals and insulating glass or double-glazed units of various thicknesses. The key to the double-glazing feasibility is that steel-window design has the advantage of lessening the size of the retaining/glazing beads. But their traditional framing system is 32mm wide with 5mm-thick flanges and is reminiscent of the elegant slimline windows and doors of the Art Nouveau period of 1890–1910 and the Art Deco period of 1920–1930s. Interestingly, the Crittall company also have a wide range of designs for domestic and industrial façades, built around purpose-made doors and windows. My thanks to the company for emailing me their SWA (Steel Window Association) detailed Guide for upgraded steel windows, is expressed here by giving their website at www.crittall.co.uk.

Crocodiling/Alligatoring: See **Alligatoring/crocodiling**

Cross-garnet hinges: These traditional-style hinges, often referred to as Tee-hinges, because of their elevational letter T shape, are still in use nowadays on ledge-and-brace style doors.

Crown rafters: These are the central rafters of a hip end roof, sometimes referred to as *pin rafters*. The rafters that diminish on either side of a crown rafter are called jack rafters.

C-studs: A modern, metal stud-partitioning component, likened to the capital letter C – because of its sectional shape – and used for wall-studs and/or door-studs.

Cup, cupped or **cupping:**   Concave or convex distortion across the face of a board, usually caused by the board's face being tangential to the growth rings in the sapwood area.

Curb rafters: A traditional reference to the rafters used for the upper, shallow-pitched slopes of a *mansard roof*. The rafters of the lower, steeply-pitched slopes are usually birdsmouthed at their top to support the timber-curb beams above, acting as wall-plates for the shallow-pitched roof.

Curtail step: This is a spiral-shaped, scroll-ended step, which was traditionally used as the bottom step of a *geometrical* staircase. It is occasionally used nowadays as the bottom step of *closed-string* staircases. In geometrical stairs, its spiral shaped end had to conform to the baluster-supported spiral-shaped, wreathed handrail-scroll above it. This is because the strings and the handrailing of geometrical staircases are continuous around numerous flights and are *not* supported by newel posts. The so-called *geometrical handrailing* – containing twisted *handrail wreaths* at junctions – is (with one exception) supported on lathe-turned balusters (referred to nowadays as *spindles*) only. The *one exception* is that there had to be an ornate, wrought-iron baluster in the centre of the curtail step, to give this end of the balustrade structural stability. This was achieved by housing the metal baluster deeply into the geometrical centre of the wooden-cored curtail step.

Curtaining: In painting and decorating, this is a trade-reference to thick-edged, wavy paint-sagging lines in a paint-film (traditionally referred to as *fat edges*). This is usually only seen on vertical surfaces and it is caused by excessive paint-flow via the overloading of a brush and/or amateurish brush-application.

Curtain-wall construction: This term refers to a relatively simplistic, less-costly and speedier form of construction which was introduced just after WWII, in about 1950. It was used on institutional and commercial buildings, including schools and local-authority houses, etc. They were built with non-load-bearing elevations on the front- and rear-aspects of the buildings, which were referred to as *curtain-walls* – meaning that they did not require foundations, as such. And this was because these sides of the buildings were wholly comprised of lightweight, purpose-made timber frames – ready to be glazed in their upper areas and clad or panelled in their lower areas. At ground-floor level, these lightweight units did not need traditional foundations – because the oversite-concrete floors were *toed* into the ground to support their minimal weight. And at first-floor level and the roof's wall plate-level, the brick structural-walls (flanking the front and rear of the building) carried the weight of the two loads imposed upon them: they provided bearings for the first-floor's joists and bearings for the roof structure. Note that the front and rear curtain-wall façades of these buildings appear to have suffered badly over the years. A variety of surveys have reported that they have been built with poor-quality softwood joinery units (which may have also suffered from poor maintenance – and are now suffering extensive wet-rot decay), poor quality cladding below the windows and poor thermal insulation within the cladding.

Curtilage: The total area of a plot of land designated for a dwelling, etc., as detailed by Land Registry and as shown on an Ordnance Survey map. Such plots relate to a 'building line'.

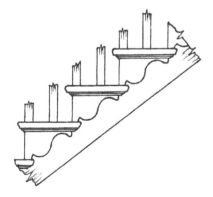

Figure 32 Cut-and-bracketed stair-string

Cut-and-bracketed stair-string: *Figure 32* (page 75):   This term refers to the *outer string-board* of a traditional *geometrical staircase*, which has triangular-shaped portions of the string's top edge removed to accommodate the visual nosed-ends of the ornate-bracketed tread-boards.

Cut-and-mitred stair-string: This is similar to that described above for a *cut-and-bracketed stair-string*, except that there are no ornately-shaped brackets embellishing the string and the riser boards are mitred to the vertical cuts of the stair-string.

Cut-tile-mitred valley (with lead-soakers):   Such a valley is formed entirely with one-and-a-half-width plain tiles (usually referred to as 'tile-and-a-half' tiles) – which are normally used at verge-edges to create the staggered plain-tiling bond. But their extra width in this type of valley allows them to be splay-cut-mitred on each alternate edge to form the mitred abutment of the cut-tiled valley. As this type of valley does not require a valley board, the tile-battens are run into the valley (to the required gauge) and are mitred on the centreline of the valley-rafter edge. Each pair of splay-cut-mitred tiles, when placed together, one above the other, in their valley-positions, will automatically form staggered outer-edges to meet the abutment of the common plain-tiles. Essentially, as the mitred valley is formed, heavy-gauge lead soakers must be interleaved and fitted between and under each pair of splay-cut-mitred tiles. Their top edges should be dressed over the tiles – and their bottom edges should not protrude.

Note that if this traditional valley was used nowadays, a modern-day breathable, impervious membrane (such as Tyvec Supro or Marley Supro) would need to be draped over the rafters, beneath the tiling battens – as well as over the whole roofing structure.

Cylinder lock (night latch):   This type of lock/latch is commonly used on traditional front-entrance doors. It consists of two main parts: **1)** a surface-fixed rim latch, fixed on the interior door-face, with a turn-knob and a sliding latch-button (for holding the latch in an open or closed (locked) position; and **2)** a cylinder with a projecting alloy-bar that fits into the latch from the exterior door-face. The *box staple* (also referred to as a *keep*) is fixed to the doorframe to receive the protruding quadrant-shaped latch. It is often referred to as a *night latch* because of its *sliding latch-button*, which can be used to lock the door internally at night (without using a key) – and can be easily slid open quickly for an emergency exit.

D

Dado rails: *Figure 33*:   Traditionally, heavily- or lightly-moulded rails were fixed to staircase- and/or room-walls at about 900mm/3 ft. above the stair-pitch line or floor-levels of the rooms. Their origins seem to have been as an ornate capping/demarcation line above the lower areas of a wall, which were invariably heavily embossed with thick

Figure 33 Ex. 75 × 25 mm Dado rail

anaglypta paper. And this was usually stained and varnished against wear and tear. Another form of dado rail was/is used as a transitional capping for wall-panelling to the lower areas of walls and stairs. Usually, the height of dado rails coincides with the overall height of chair-tops – seemingly to protect the wall's décor.

Dado tiles: Traditionally, dado ceramic-tiles with protruding, moulded shapes similar to dado rails, were used as an ornate capping to half-storey-height tiled walls in bathrooms and WCs. In this era, the tiles were bedded individually with sand-and-cement mortar, resulting in a tile-and-mortar-bed thickness of 18mm/ ¾in. Alternatively, edge-tiles with one or two quadrant-shaped edges were available. Nowadays, edge finishes at top- or reveal-edges seem to be predominantly in the form of prefixed metal and/or plastic trim.

Damper plate: Traditionally, an adjustable/sliding metal plate was sometimes built into (across) a chimney flue, for the purpose of regulating the down-draught.

Damp-proof courses: See **DPC defects and remedial treatment**

Dancing steps: Traditionally, these were consecutive steps with tapered treads which did not radiate from a common centre point; they radiated from eccentric points beyond the newel post, which makes the narrowest side of the treads/steps wider – and therefore safer – than those that radiate from a common centre point. However, dancing steps would present a greater challenge to designers regarding the present-day Regulations concerning the 2R+G formula around double pitch-lines.

Datum: A fixed and reliable reference point from which all levels or measurements are taken, to avoid cumulative errors.

Daylight window-sizes: From a room-size point-of-view, this is usually calculated as being at least 20% of the intended room-size.

Dead knot: This term refers to a knot which is loose, broken or decayed; an ominous black ring around a knot is usually a sign that it is likely to fall out when the moisture content of the timber changes.

Deadlocks: See **Mortise Deadlocks**

Dead shore: *Figure 34*:   This term is used to describe a method of temporary support to a load-bearing wall, whilst the wall is being partly removed to enable a beam to be inserted and an opening created below it, for a doorway, etc. Initially, one or more small holes are made through the wall, slightly higher than the intended opening, and temporary timber-beams (referred to as *needles*) are centrally positioned through them. These needles, in turn, are then supported near each of their ends by telescopic metal-props (which were traditionally timber props, known as dead shores). The props, whether of metal or timber, should be temporarily fixed to timber *sole plates*. Once the opening has been cut and formed below the needles (which might include forming piers and/or padstones on each side), the beam (of steel nowadays) is

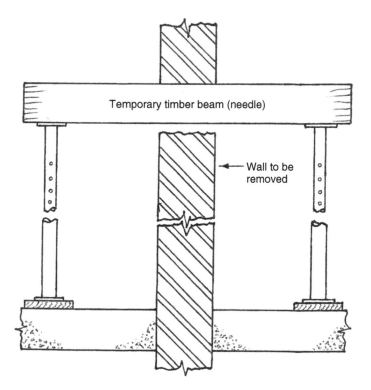

Figure 34 A modern form of 'dead shore' created by using telescopic steel-props on timber sole plates

inserted. Then the underside of the brickwork above its top flange is *made good*. After this has set, the needles and dead shores, etc., can be removed – and the needle-holes can be made good.

Deadwood: This refers to wood from a dead-standing tree that lacks strength and its usual weight, symptomatic of trees that are felled well after reaching maturity.

Death-watch beetles: These insects, of the *Xestobium rufovillosum* genus, are similar to the furniture beetles, but are much larger (being 6 to 9mm/¼ to ⅜ in. long) and their flight holes are larger, about 3 to 4mm diameter. The beetles' larvae burrows into the sapwood areas of structural timbers, seemingly with an attraction to English oak – and historic laboratory research suggests that these beetles seem to be predisposed to oak that is already under attack from a fungus known as *Phellinus megaloporus*. Concaved, honeycomb-pitted damage is often found in the rafters and beams of old buildings and churches. The beetles' name derives from their historic attraction to church roofs in relation to the ticking/tapping noise that the adult beetles make during the mating season. In an old, oak-framed dwelling house in the Weald of Sussex, where death-watch beetles were found to be active (about a decade ago), the author was given a personal account of how the tapping noises at night kept the two elderly occupants awake.

Treatment to eradicate such infestation and to assess and repair any major structural damage is usually best done by specialist companies such as Rentokil, who can usually give an assurance and a transferable, written guarantee for a period of years.

Decimetre: One tenth of a metre, i.e., 100 millimetres/10 centimetres (100mm/10cm).

Deck/decking: A trade-reference to structural floor-coverings, such as boards or sheet-material, i.e., T&G floorboards, chipboard, or plywood, etc.

Deliquescent soot-ash staining to chimney-breast plaster: Black, sponge-patterned stains can mysteriously appear on the plastered and decorated surfaces of chimney-breast walls in dwellings of yesteryear, built prior to the Codes of Practice in the 1965 edition of the Building Regulations. However, since the implementation of this publication's criteria, the staining problems caused by brick-formed, parged chimney-flues have been solved by using separate, flush-jointed and smooth-surfaced flue-linings that are built-in to the rough brick-flue structure around them.

But the problems of staining to older properties' chimney-breasts still needs to be understood and addressed, so is herewith explained:

The historic, rough method of *parging* (plastering) the brick flues– as described under the **Parging** (or *pargeting*) entry in this book – with soft, deleterious mortar (which contained undissolved pockets of hydrated lime), gave a fair service whilst the flues were in use with open fires in their hearths, but, after a period of non-use via the cessation of open-fires, the staining problems began to manifest themselves. And it was eventually realized that the problems were related to the *disuse* of the old flues, coupled with the *sealing-up* of fireplace-openings and, in some cases, the partial removal and *capping* of the above-roof chimney stacks – exacerbated in other instances by the *non-provision* of *air-vent grilles at each sealed opening* and *to the uppermost portion of the capped chimney-stack*.

Such change of usage and ventilation omissions cause a flue's retained soot-ash deposits and sulphates to become wet via condensation, changing them into a distilled, deliquescent liquor that may gradually leach out through the flue's ½-brick-thick walls, staining the plastered and decorated surfaces with black, sponge-patterned stains.

Remedial treatment could be attempted with Stain-Block paint in aerosol or brush-on form, prior to redecoration – but, if the deliquescence-staining still bleeds through, more drastic action will be needed in the form of plaster-removal and relining the wall(s) with a modern-day dry-lining. This could be a version of the established technique of lining walls with sheets of plasterboard by a '*dot-and-dab*' technique. However, the *dabs* of fixing-plaster (normally between the wall and the plasterboard, in between the *dots*) could not be used on the contaminated areas of wall. But the *dots*, comprised of a number of small pieces (say, 75mm x 75mm) of 12mm-thick bitumen-impregnated fibreboard, could be bonded to the wall(s) with fixing plaster. The other variation from the norm would be that the fixing of the plasterboard would have to be done to the dots – not the wall – via the application of a good-gripping panel adhesive.

Descending damp (via chimney-stacks, raised party-walls, or window-sills, etc.): Yesteryear's lime-mortar joints, if subjected to excessive moisture-absorption via the deterioration of external joint-pointing to chimney stacks, or to raised, fire-break party walls' pointing or rendering, can cause the lime-mortar of the brickwork below the roof (in the roof-void) to become decomposed and friable (powdery), making such brickwork structurally weak. And any nearby trimmed- or trimming-rafters around such stacks – or any common rafters too near the party walls – are likely to be stained with white efflorescence and affected by wet-rot decay. If the brickwork of gable walls and chimney breasts (or independent chimneys) within a roof-void (loft) indicate high moisture-meter readings, this is strong evidence of descending damp from the chimney stack(s) and/or the raised party-walls above the roof. And to confirm whether the mortar joints are decomposed, a few bed-joints should be pierced with a sharp-pointed tool such as a bradawl. If the mortar has decomposed, the bradawl will sink in quite easily – often up to the hilt. If further confirmation is needed, a light raking out of mortar in a small, trial area should produce decomposed mortar in the form of a very fine, powdery sand.

Detached and **link-detached houses:**   The former is self-explanatory, i.e., the property is separated from other properties (be it only by a miniscule amount nowadays) – and the latter infers that some part of the property is conjoined, usually by a shared party wall between two garages, preceded by a shared driveway.

Diagonal-cover slating: See **Asbestos-cement roof-slates**

Diamond-cover slating: See **Asbestos-cement roof-slates**

Differential expansion: This term refers to the different coefficients of *thermal expansion* that exist and take place between different materials that are used together in a material-mixed construction. And it must be realized that these different rates of expansion in closely- mixed materials may cause structural damage. For example, a 610mm/2 ft. width of plastic-laminate (such as Formica), bonded (glued) to the top-side only of a laminated-timber (gluelam) kitchen worktop, would very soon suffer from *differential expansion* via the thermal movement in the worktop conflicting with the different rate of thermal movement in the plastic-laminate – and serious *cupping* would take place. To have prevented this, in this particular instance, a balancing veneer/membrane of plastic-laminate was required to be bonded onto the underside of the worktop.

A second example of differential expansion concerns a number of precast, imitation-stone window-sills that, after a decade of being bedded on the apron walls of a building's window openings, developed a crazed pattern of fine surface-cracks. They were monitored for quite a period of time, but did not deteriorate further – and still seemed to be weather-tight. So, it was concluded that differential expansion had taken place between the main body of the precast material (the reinforced concrete) and the top- and side-areas of the artificial stone (which is usually comprised of crushed stone, selected sand and white cement). The analysis could not be confirmed, because the building's owner decided not to disturb the sills for an invasive investigation.

A third and final example of differential expansion is related to a vitreous-china wash- hand basin, which was only about five years old. Its white glaze inside and outside of the bowl was extensively crazed with fine, irregular cracks. But it did not leak – so it was concluded that differential expansion (thermal movement) had taken place between the white glaze and the vitreous-china basin, via a manufacturing fluxing fault.

Dihedral angle: The angle produced between two surfaces, or geometric planes, at the point where they meet. For example, two vertical surfaces that meet at right-angles to each other, produce a dihedral angle of 90°, but two *inclined* surfaces that meet at right-angles to each other – such as a hipped roof or valley formation – produce a dihedral angle other than 90°, according to the degree of inclination.

Dilapidations Report: See **Schedule of Condition** or **Dilapidations Report**

Dimensional coordination: This refers to building-components that are made – by agreement between relevant parties – to fit into a modular system.

Dimensional stability: This refers to the degree that building materials suffer movement which affects their size and performance with regard to thermal expansion and contraction.

Diminishing stile/gunstock joint: Traditionally, to give a half-glazed door more glass-width above the mid-height lock rail, the stiles in the glazed area were reduced in width to those below – in the panelled area – resulting in splayed/diminishing stile-joints which were thought to give the stiles the appearance of a long rifle: hence the name *gunstock joint*.

Direct- and indirect-heating: Direct heating refers to a room being heated by a source of heat such as a gas-, electric- or woodburning-fire (or stove) within it – whereby *indirect-heating* refers to the heating of a room or rooms from somewhere outside of an individual room, and being brought in by hot water or hot air via gas- or oil-fired boilers, or an electric heating unit. These are commonly referred to as **central heating systems**.

Direct hot-water supply system: This description refers to a centralized method of hot water supply, whereby all the heated water drawn off the appliances has passed through the boiler. See also: **Indirect hot-water supply system**, written herein.

Distribution board: See **Consumer unit**

Dogleg (or doglegged) staircase: This refers to two short-flights of stairs, between a single storey-height, which are connected by a half-turn landing (a landing that changes the flight-direction by 180^0). The historic *dogleg* terminology came about because the outer string- board's top tenons of the lower flight were mortised-and-tenoned into the same side of the landing's newel post, as the other, upper outer-string board's bottom tenons, which were located immediately above those of the lower string. This connection, with both strings going in different directions in the same plane, was thought to look like a bent dog's leg: hence the name. Although there are many doglegged stairs still around, they do not meet present-day Building Regulations. This is because the bottom flight's handrailing – via the dogleg feature – is not/cannot be continuous. However, changes in Building Regulations cannot be enforced retrospectively. Note that this type of staircase, in present-day design, usually meets the new regulation by providing two newel posts, side-by-side, at the landing's change of direction.

Dogtooth course: An ornamental *string course* feature, formed in brickwork by laying a course of *half-length bricks* diagonally at 45^0, so that the outer-corners of the sawtooth-pattern feature are flush to the face of the wall.

Door-base weather boards: See **Weather boards**

Door case/casing: An archaic reference to a door-lining.

Door closers: Spring- or hydraulic-fittings are used to enable an opened door to self-close automatically. They can be at the foot or the head of a door – or in its centre-edge area. They are usually required by the Building Regulations Part B, Fire Safety, to be fitted to certain designated doors.

Door frames: These are usually for housing exterior doors and should, therefore be of sturdier construction than a door lining. And, unlike interior linings, that have loose doorstops, frames should be rebated to receive the door and usually have hardwood sills.

Door furniture: This second-fixing carpentry term refers to lever- and/or doorknob-handles, locks and latches, hinges and bolts, etc.

Door joints: These are the necessary gaps (called *joints*) of 2mm to 3mm around the edges of a door for opening clearances.

Door-linings: 1) Wooden: These three-sided *casings* should, by definition, completely cover the reveals (sides) and soffit (underside of the lintel) of an opening, as well as support and house the door. Nowadays, this historic coverage of *all* doorway reveals only usually occurs where the openings are within a block- or stud-partition wall. Linings should not be less than ex 25mm/1 in. thick (thereby with a planed-finish thickness of approx. 20mm/¾ in.), and are not usually more than ex 32mm/1¼ in. thick (thereby with a planed-finish thickness of approx. 28mm/1⅛ in.). If being used for doorway openings in partition walls (as opposed to doorway openings in thicker, structural walls), their width must equate to the partition's thickness + the estimated finished-thickness of the wall's linings on each side.

2) Steel: In the post-WWII period of the 1960s/70s, preformed steel door-linings started being made in factories and used in house-building. They were marketed as pre-assembled units, that were complete with fitted loose-pin hinges, a striking-plate for a door latch, round-edged, plain-face architrave and door-stop projections – and backside recesses (between the architrave projections), for receiving 100mm/4 in.-thick blockwork and plastered partition-walls. It was thought that they would speed up the building process of the blockwork walls (by having non-flexible, pre-positioned door-linings, acting as plumb, stabilizing profiles); whilst reducing the slow, first-fixing carpentry operation of fixing wooden door linings. But these preformed-steel door-linings did not open any doors for the designers or manufacturers. They were not liked. They were cold to the eye and to the touch. And the preformed rebates were oversized in relation to the door-thickness – and carpenters never seemed to bother about offsetting the hinge-knuckles to compensate for this. Apart from all that, small (9mm/⅜ in. diameter) door-cushion pads/domes had to be glued to the steel-edges of the protruding doorstops – and renewed too frequently.

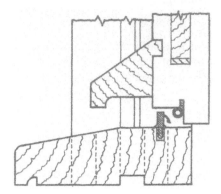

Figure 35 140 × 45 mm hardwood door-sill with a weather-slope, anti-capillary groove and a PVC water-bar

Door-seals: See **Draught excluders** and **door-seals**

Door sills: *Figure 35:* For purposes of weathering, draught exclusion and structural transition between exterior and interior paving and floor levels, external doorframes usually have hardwood sills – sometimes referred to as *thresholds*. Traditionally, *water bars*, with a sectional size of 25mm x 6mm/1 in. x 0¼ in., ran along a groove in the top of the sill, protruding by 12mm/½ in. to form a water check/draught excluder. These bars were made of brass or galvanized steel, but when this form of weathering is used nowadays, the bar is available in grey nylon or brown-coloured rigid plastic with a flexible face-side strip. This acts as a draught seal against the rebated bottom-edge of the door. In this arrangement, a sloped weatherboard should be fitted to the bottom face of the door.

Door stiles: See *Figure 12* on page 26: The two vertical components that occupy the outer edges of a framed door, usually of timber of either a 45mm/1¾ in. finished thickness for an outer door, or of 35mm/1⅜ in. finished thickness for an internal door.

Doorstops: As implied, these are to *stop* a door when it reaches the closed position within a *door-lining* or a *doorframe*. However, interior, wooden door-linings have separate doorstops fitted-and-fixed to their head and sides (referred to as *loose* or *planted* stops), but exterior-type doorframes should have their doorstops rebated from their solid head-and-sides. And the latter is technically and traditionally referred to as *sunken* or *stuck* rebates.

Dormer cheeks: These right-angled, triangular sides, traditionally formed with diagonal boarding, parallel to the roof-pitch (thereby acting as diagonal frame-supports), are nowadays usually formed with sheet material, either exterior plywood, water-resistant chipboard or Sterling OSB (orientated strand board) – which effectively does the job speedier.

Dormer windows: These are windows that are positioned vertically on the sloping surfaces of pitched roofs. They have vertical, triangular cheeks at their sides and their own, separate roofs. The latter may be *flat* (more common nowadays), semi-circular, segmental, pitched, gable-ended, hip-ended, hip-gablet ended, or of an *eyebrow* shape (see *Eyebrow windows*).

Dot-and-dab dry-linings: This term refers to an early form of plastering technique used for the dry-lining of solid walls, whereby the wall should be trued up with small, square pieces of bituminized fibreboard (referred to as *dots*) being bedded strategically onto the solid wall with a *buttering* of plaster, to *true-up* the vertical- and horizontal-alignment of the structural wall-surface, before much larger *dabs* of plaster are applied to the wall (between the dots) to receive and adhere the plasterboard-linings placed up against the cavity-creating dots. In reality, the dots seem to be avoided nowadays and the truing-up of the boards appears to be done by applied-pressure-judgement, related to a long spirit level instead.

Double-acting hinges: If not referring to more sophisticated hinges, that swing a door through 180^0 or more, this description refers to **helical hinges** (also referred to as *Bomber butts*, by virtue of their appearance being thought to resemble a pair of high-explosive bombs – no doubt because of the popularity of helical hinges just after WWII).

Double angle-bead in plaster: This traditional *double-bead* shape – incorporating a shallow quirk (groove) feature, inset from each side of a semi-circular-shaped edge-mould-bead – was used in pre-20th-century properties as an ornate/protective angle-bead. It was formed on the corners (quoins) of interior, rendered-and-set plastered walls and on the corners of projecting archway piers, etc. Their actual purpose was to protect the arrises of the wall – so they were usually formed with strong *keens'* cement prior to the plastering of the actual walls.

Double doors: This expression refers to a pair of doors that each swing separately, from their closed, central meeting-position, to one side only (being further described as *double doors with a single swing*); or that swing separately to *either* side of their closed position (being further described as *double doors with a double swing*).

Double-faced architraves: Sometimes, in second-fixing carpentry work, door linings are fixed on each side of the end of a partition wall (when two doorways are adjacent to each other), and a double-faced architrave is used – as a better alternative to leaving a very small gap between two separate, back-to-back single-faced architraves.

Double floors: *Figure 36*: This is usually a technical description that relates to a single floor for its explanation; so, a suspended timber-joisted floor consists of thick, board-like timbers (joists) placed on edge and spaced parallel to each other at specified centres (usually 400mm/16 in. or 600mm/23⅜ in.) across a floor and, in the case of a *single floor*, resting between the extreme bearing points of the walls. But in the case of a *double floor*,

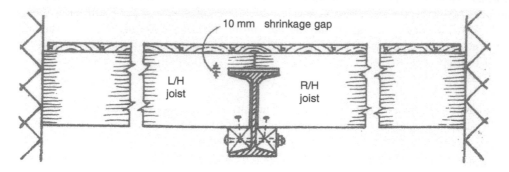

Figure 36 A traditional 'double floor,' with a central RSJ (rolled-steel joist) acting as a mid-area cross-beam

they rest on one or more intermediate supports (at right-angles to the joists) and the extreme bearing points of the walls.

Double glazing and **triple glazing:**   As these terms imply, such window- and door-panes (referred to as *units*) are comprised of two or three panes of glass – and this is done to achieve lower U values and increased thermal efficiency. The glass panels are separated around their edges by so-called *spacer bars* (manufactured with insulating features), which are produced in varying thicknesses from 6mm/¼ in. to 20mm/¾ in. The spacer bars *seal* in the air at low pressure (it does not create a vacuum). And because air naturally contains argon gas, this acts as an insulator to prevent the warm air escaping from a dwelling.

Triple glazing simply contains three panes of glass, separated by two separate, perimeter-edging strips of insulated, *warm-edge* spacer bars. If required, the sealed space between the panes of glass can be filled with a gas to increase the unit's thermal performance. Also, low-emissive (Low E) glass, which keeps more heat in, can be used or specified. Finally, double glazing has a U value of about 1.8, compared to a lower U value of about 0.8 for triple glazing.

Double-hung sash windows: See *Figure 13* on page 30: This refers to a pair of vertically-sliding sash-windows housed in so-called boxframes (derived from *boxed frames*). The weight of the sash-corded sliding windows is counterbalanced by lead- or cast-iron weights housed in the boxframes, allowing the outer sash to be slid down and the inner sash to be raised up. Still very much in evidence in the UK, they date back to the late-17th century.

Double-margin door: This refers to a single, traditional-style panelled-door, made to look like a pair of doors, i.e., like *double doors*. They were usually made with hardwood and can still be found gracing the entrances of libraries and public buildings, etc., even though the former may not, nowadays, contain books any more. The door consisted of one double-length top rail, four stiles, two bottom-rails, two lock-rails, and (usually) four intermediate rails (two for each side of the door's central stiles). The two outer stiles were mortised and tenoned to the various rails in the usual way – as were the two inner stiles, with the exception that where the inner, meeting-stiles were joined to the one-piece top rail, they were bridle-jointed to its centre – giving the visual impression of two separate end-of-top-rail mortise-and-tenon joints. Other final features included a recessed metal bar on the underside of the door, screwed in across the separate meeting stiles with brass screws – and pairs of secreted wedges glued and driven across corresponding mortises in the meeting-stiles' edges, prior to panel-insertion.

Double-skin or **double-leaf walls:**   See **Cavity walls**

Dovetail key: Apart from referring to the locking effect of a dovetailed shape in joinery or masonry, etc., this also refers to nails driven in at opposing angles to each other to create a dovetailed effect between two pieces of timber – such as at the joint between a door-lining head and a door-lining leg (although these are commonly screwed together nowadays).

Downpipe: See **Rainwater pipe** (**RWPs** or **downpipes**)

DPC defects and remedial treatment: **1)** Traditional DPCs (**damp-proof courses**) listed above as being of single- or double-layers of quarried slates or clay tiles, are likely – because of their inflexibility – to suffer hairline cracks across their width via structural movement of the surrounding brickwork. And such cracks are extremely likely to cause rising dampness via capillarity/capillary attraction. **2)** Traditional bituminous-felt DPCs, of approx. 3mm/⅛ in. thickness, are flexible enough to withstand minimal structural movement; their weaknesses usually relate to any undersized overlapping at their lengthening joints. This should not be less than 100mm/4 in. at right-angled joints and not less than 150mm/6 in. at lengthening joints, to combat horizontal capillary attraction between the layers. And, importantly, the overlapping area at all joints must be completely free of mortar contamination. Note that horizontal capillarity at DPC joints eventually manifests itself as rising damp.

Remedial treatment via liquid-chemical injection, or pelleted chemical-plug insertion, into closely-drilled holes in the base areas of affected walls are two of a range of treatments that can be carried out. There are also repetitive hit-and-miss DPC-replacement procedures (that sound good, if done properly) – or there is the removal of interior solid-plaster surfaces up to 1m/39⅜ in. above floor levels, followed by replastering with sand-and-cement rendering that contains a waterproofing additive.

DPCs: This common abbreviation for **damp-proof courses**, refers to **1)** an impervious material such as the traditional form of single- or double-layers of quarried slates or clay tiles; **2)** bituminous felt (introduced in the early 1930s); or **3)** brick-width black-plastic DPCs (from about the 1980s), bedded horizontally in the foundation brickwork at not-less-than 150mm/6 in. above ground- or paving-level. DPCs are to inhibit *rising* *damp*, caused by capillarity (*capillary attraction*) and to lessen the risk of rain splashing-up to above the DPC from any paved surfaces against a wall. Note that if the slates or clay tiles mentioned above are of a double layer, the header-joints of the top layer of slates or tiles ought to be centrally staggered in relation to the joints below. Modern plastic DPCs are manufactured with a raised criss-cross pattern on both surfaces, which is essential to ensure that a physical attachment will develop when sandwiched between the top and bottom layers of completely-set mortar-beds. Such an attachment is necessary to inhibit lateral movement (sliding) of a brick wall above DPCs as a result of ubiquitous thermal movement.

DPCs are also used at other vulnerable parts of a building, such as **a)** above door- and window-openings in outer walls; **b)** where a parapet wall connects with a roof; **c)** where a chimney meets a roof – if a *lead-safe* is not installed; **d)** under wooden window-sills; **e)** under stone or precast-concrete sills, especially if mid-area butt joints are used.

Another form of DPC is also essential at the window- and doorway-reveal abutments that link/close up the open cavities with (nowadays) preformed cavity-closers.

DPM: Similar to DPC, this trade abbreviation is for *damp-proof membrane*, referring to an impervious membrane such as heavy-gauge polythene, bituminous liquids or asphalt, etc., covering (or sandwiched between) the whole of the concrete ground-floor slab, below the sand-and-cement floor screed – or below the insulated floor, to inhibit *rising damp*.

Dragon beams: 1) These timber beams were used in 15th- to 17th-century buildings that were designed with so-called *jettied* façades, whereby the ground-floor of the front elevation (and sometimes the side elevations) *receded back* from the building-line (face) of the floor, or floors above, by at least 300mm/12 in. Or, it could be described as: *the façade(s) of the upper floor(s) above the ground-floor elevation(s), protruded forward from the building-façades below*. In practical terms, when the side elevations were jettied, dragon beams were set up to protrude at 45° beyond the front corners of the building – and they were mortised to receive 45° oblique-tenoned, jettied floor-joists, set up at right-angles to each other on each side of the beams. The dragon beam itself was only about 2.130m/7ft. long and relied on an angled tusk-tenon joint into the side of a bulky main-beam that spanned across the building.

2) In the 18th to 19th centuries, *dragon beams*, in a much-modified form, were used in hip-ended roofs, at the base of hip rafters and connected to so-called *angle ties*, which were fixed diagonally across the right-angled wall plates. The dragon beam housed the end of the hip rafter and was tusk-tenoned to the side of the angle tie to prevent undue tensile stress being transferred from the hip to the corner of the roof-structure.

Dragon ties: See **Angle ties**

Drain cock: A draining-off *cock* (water outlet), which is placed at the lowest point of a water system to allow the system to be drained, when necessary (for repairs, etc.), usually via a hosepipe attachment.

Drain-condition Report: During a Building Survey, surveyors do (or should) inspect accessible parts of a property's drainage system – which should include the condition of the gutters, gullies and rainwater downpipes, if only via binoculars – and whether the number of RWPs seems adequate in relation to the various lengths of guttering. It should also include the visual condition of the interiors of all inspection chambers (manholes) discovered on the site – but only if the oblong- or circular-shaped covers can be lifted. Occasionally, surveyors find a profusion of fine tree-roots on the brick-wall surfaces of traditional inspection chambers, where the mortar joints have eroded and opened up to allow such intrusion. In their search for water, these roots can also be seen partially blocking the orifice of the exit pipe at invert level in the chamber's wall. One also comes across blocked or partially blocked drains, that should be verbally reported ASAP, as well as being written up in a Report. Finally – when traditional chambers are not blocked or partly blocked – the semi-circular SGW (salt-glazed earthenware) channel in the chamber's mid-area and the open-ended branch-pipe bends, plus the sculpted sharp-sand-and-cement *benching* at perimeter edges, all need to be surveyed for cracks-and-fissure damage.

Also, any evidence of soil waste (excrement and/or toilet paper fragments) deposited on the chamber's walls, or on the upper contours of the perimeter benching, must also be reported upon – as a sign of periodic chamber-blockage, needing further investigation by a drain specialist.

Drain pipes: Modern underground drainage uses 110mm/4⅜ in. diameter uPVC pipes and fittings, with ring-seal joints for both expansion and sealing, to BS EN 1401-1: 1998 and BS 4660. And polypropylene inspection/access chambers and gully traps, etc., have superseded brick-built inspection chambers and SGW (salt-glazed earthenware) drain pipes. Although, these traditional products are still available for drainage conversions and/or repairs.

Drain-pipe Survey: This is carried out by specialist drain-repair contractors and usually involves the use of a CCTV (closed circuit television) camera. Such action may be necessary if a drainage system is found to be frequently blocked and *backing up* into the shallowest inspection chamber (manhole); thereby suggesting that the pipe-run may be damaged or – if the system is of glazed earthenware pipes – that fine, tubular tree-roots may have broken through the sand-and-cement caulked joints and are impeding the flow. However, if the last, deepest chamber on the property (the penultimate chamber that precedes entry into the public sewer) is retaining liquid effluent, this suggests that the intercepting trap into the sewer is blocked and needs clearing– and that a CCTV survey may not be necessary.

Drain-relining: See **Relining drains**

Draught excluders and **door-seals:** There is a wide variety of excluders and seals available nowadays, that require at least 3mm/⅛ in. wide grooves to be made to varying depths in, or adjacent to, the rebated window- or door-stop edges, for the excluders or seals to be held by their friction-fit fins. Unless you have sophisticated woodworking machinery to form such grooves, a router cutter will be required – and an electric router, of course. The core information given here has been gleaned from a catalogue of *IronmongeryDirect.com*; Telephone 08081 682828. Apart from their ironmongery, they have a wide selection of excluders and seals. The AQ21 AquaMac Seals, that conform to BS644/BBA, are reckoned to give a high performance for windows and doors – and, to my certain knowledge, they do.

Drawbore dowels: This refers to the traditional and very effective method used for cramping up oblique mortise-and-tenon joints used on stair-strings and handrails – which cannot be easily cramped by other means. Hand-tapered wooden dowels are driven into offset holes drilled (separately) though the newel posts and the string or handrail mortises. The offset holes in the tenons effectively pull up the shoulders when the dowels are driven in, which permanently reinforces the glued joints.

Drip-grooves: See *Figure 3* on page 6: These are small grooves under the projecting outer-edges of coping stones, window- or door-sills, etc., to inhibit rainwater from creeping/flowing back (by capillary action), or being driven back by wind towards the structure of the building.

Drip moulds: See *Figure 17* on page 42: Traditional, wooden casement windows (side- and top-opening) and BWMA (British Wood-machining Manufacturers Association) casement windows used 50mm x 40mm and 32mm x 32mm weathered drip-moulds respectively, over their uppermost windows.

Drones used in land- and building-surveying: Drones are *unmanned aerial vehicles* (UAVs), which are controlled remotely from the ground by trained individuals or specialist companies. Their function in *land*-surveying, prior to architectural planning and design, is via the use of a laser scanner. This makes a photographic record of a ground-area's surface, which is later enhanced via visualization and modelling software and is used to produce a three-dimensional image of the area. This modern practice is more accurate and faster than conventional site-engineering/land-surveying measurement and triangulation methods.

When used for *building*-surveying, a roof or any obscure part of a building can be analysed – if necessary – thus producing photographic images via their downward-facing sensors and cameras.

It is believed that drones cannot be used to fly over military bases, airports or prisons.

Note also that the Drone (Regulation) Bill 2017–2019 differentiates between domestic drones and commercial drones. If the owner of a drone is being paid for its use, then a licence is required.

Dry-Fix Roofing Systems: Partly due to the unreliability of the adhesive qualities of mortar-bedded hip- and ridge-tiles – and partly due to a time-saving advancement to lose a wet-trade operation, *dry hip- and ridge-systems* using patented plastic components, aided by underlying, adhesive *ridge-* and *hip-rolls* of flashing material, *hip trays* and *dry valley-liner systems* are being used on newly-tiled roofs.

Dry-lined walls: See **Dot-and-dab dry-lining**

Dry rot: This contradictorily-named wood-destroying fungus is a *moisture*-dependent, malignant disease, botanically named *Serpula lacrymans* (that was historically named ***Merulius lacrymans***). It is related to the mushroom family – and is also known as a **brown rot**. It can take over from **cellar rot**, as it needs damp, enclosed locations to thrive – especially where there are humid conditions, timbers with above 20% mc (moisture content) and poor ventilation. The *dry*-rot contradiction, however, is a good description of the timber's resultant low moisture-content and surface-damage, as softwood that has suffered an attack by this fungus is left relatively dry and gives the appearance of having been burnt or baked to a dull brown colour, with deep, criss-cross shrinkage-cracks on its surfaces in the form of conjoined cube-like fissures which run with and across the grain. At this stage, the timber has lost its strength and – depending on its location as a floor- or ceiling-joist, etc. – is likely to collapse. If tested with a penknife, little resistance will be met; and it should be possible to easily dig out a portion of such affected timber, which could then be crumbled to dust quite easily between the forefingers and thumb to confirm that the aftermath of dry rot has been discovered.

The cycle of dry-rot development starts with destructive mushroom-like *sporophores*, also referred to as *fruiting bodies*, that initially form on diseased, dying or fallen trees – nature's way of biological disposal. The fruiting bodies (in the form of *stalks, stems, brackets* and *plates*) develop many millions of *spores* (seeds) of microscopic size (1/100th of a millimetre) and of measurably-unrecognisable lightness of weight. Once the spores are released naturally, after their spore-bearing surface – *hymenium* (hymen in human biology) – has broken, or they are disturbed unnaturally (by wind, etc.), they are released and become airborne. And they are reportedly capable of remaining suspended in the air for a very long time, drifting over considerable distances. The relatively small percentage of them that touch down on fertile (i.e., damp) wood within a building may trigger a dry-rot attack if the wood has a moisture content in excess of 20% mc (known as the *dry-rot barrier*) and which also happens to be situated in a position of poor ventilation, high humidity and relative warmth between 23°C and 25°C. In such conditions, the spores are very likely to germinate, after turning into visually-noticeable surface-spots. And their roots, *hyphae*, will develop into *mycelium* fungus. Once established, it will spread through the body of the timber, consuming the cellulose from the over-moist, woody tissue of the sapwood and less-durable heartwood. Eventually, the mycelium develops on the diseased timber's surface in the form of a mass of tubular, moisture-carrying threads. During this process, second-generation fruiting bodies take growth on the diseased timber in the form of *plates* or *brackets*. Still seeking more food, the developed tubular-strands of the mycelium, having increased in size up to about 6mm Ø and concealed within a white, cotton-wool-like blanket, will spread out in search of fresh supplies of wood.

This search is relentless and can start in a cellar and end up in the roof void, after attacking the timber floors in between. Or, vice versa, it can start in the roof void and end up in a cellar – or the timber-suspended ground floor. Adjoining properties, via party-walls, are also at risk. This is because the mycelium's moisture-carrying thread-like tubular-strands are capable of insidiously working their way through any fine gaps in the mortar joints of brick walls and of spreading over the surface of walls, in between the wall and the traditional, solid plaster. Older buildings are at greater risk, because untreated joist-ends, timber wall-plates, beam-ends and skirting boards, etc., are usually up against single-skin, potentially-damp brick walls and solid plaster – making them likely to be wet enough to be susceptible to the dry-rot attack.

To give a word-picture example of the unmistakable signs of an attack of dry rot to a timber-joisted floor, suspended on low-rise sleeper walls above an unmade oversite, one might see (once a number of the obviously-damaged floorboards had been removed) the dry-rot mycelium in the form of a white to pale-grey coloured blanket of finely-matted, thick fluffy cobwebs draping over and bridging across the affected joists – and covering surface-areas of the oversite. Once exposed to light, the edges of the mycelium blanket sometimes change to lilac and/or yellow.

Finally, the signs of an attack are: **1)** an unpleasant, *musty, mushroom-like odour*; **2)** partially collapsed floorboards; **3)** distorted surfaces of skirting boards, etc., that are warped and wavy (partly concaved or convex across their width) and/or are with mid-area shrinkage cracks along and/or across the grain; **4)** the possible surface-appearance of fruiting bodies in the form of pancake-like plates or brackets; note that these relatively-thin mushroom- or omelette-like formations vary in size from a ragged 20mm width/diameter, to 1m or more and are white-edged with a rusty-red centre (the red being the spore dust). On painted surfaces – such as skirting boards – that show signs of corrugated distortion running along the grain, testing with a pen-knife will likely meet with little or no resistance. Finally, note that *brackets* get their name when the fruiting bodies form in the right-angled areas of doorframe-heads and window frame-heads – or similar right-angled areas. And *plates* get their name from the fruiting bodies that form on horizontal surfaces such as floorboards – and are at least of tea- or dinner-*plate* size.

Dry-rot treatment: Traditionally, the recommended treatment of – for example – a timber-joisted ground-floor, suspended on low-rise sleeper walls, was to remove all the obviously-affected skirting boards, floorboards, joists and timber wall-plates bedded on the sleeper walls and burn them – if possible – in a safe, outside area on the same site. Any remaining, questionable timber was 'sounded' by tapping/banging with a hammer and/or by checking its resistance to pressured knife- or bradawl-piercing. The hammer technique was more common and, with minimal experience, one could approximately distinguish where the interior rot ended (by the diminished *unhealthy* dull sound giving way to the enhanced, *healthy* sharp sound). When this point was determined, one was then advised to measure 450mm/18 in. to 600mm/24 in. beyond it and cut off and destroy (burn) any questionable portions of timber.

Of course, nowadays, with a higher ratio of labour charges, related to material prices, it would no doubt be cheaper to remove the entire floor and renew it with preservative-treated timbers.

Either way, the next step before partial- or complete timber-replacement, was to check the lower walls and the underfloor oversite for fungal contamination. All brick-wall surfaces between the oversite (which may be of damp soil in an old building) and the mid-area of the removed skirting boards (where, traditionally, the plastered walls ended) are very likely to be affected with threadlike fungal strands – which may also be under/between the plaster and the wall. In such cases, the plaster has to be removed. Traditionally, such affected brickwork was tediously scorched *clean* with a blowlamp – but this technique was questionable in its effect and dangerous in its application. Nowadays, such fungal strands should be brushed off before such brickwork is sterilized and treated with an antiseptic solution.

However, bearing in mind the Health and Safety at Work Act (HASAWA) and the Control of Substances Hazardous to Health Act (COSHH), in the use of such dangerous substances, this treatment should be carried out by specialists – especially if they can provide a seemingly reliable written guarantee.

Dry-stone walls or **dry walling:**   These are so-called *Rubble walls* built without mortar joints.

Dual-flush cistern: See **Water closet (WC)**

Dubbing out: 1) Filling in any deep hollows or irregularities in a wall surface with undercoat plaster, prior to applying the overall undercoat and finishing plaster; **2)** Building up the core-shape of a traditional in-situ built cornice by *dubbing out* the wall-and-ceiling junction with plaster, prior to *running* the *running cornice-mould frame* to produce the cornice shape.

E

Earth connection: A common reference to a wired connection between an electrical circuit in a property and the earth (the ground), usually via an earth clamp to a gas- or water-mains pipe at ground level.

Easement: 1) In conveyancing law, regarding so-called onerous covenants, etc., this term refers to a legal right that a person or persons may have over another person's land, such as allowing them to walk over it, or to run pipes below-the-surface of it. **2)** In general-building terms, easement usually refers to the slackening of *folding-wedges* beneath traditional **underpinning** *props* or **arch centres**, immediately after the bricklaying work has been completed, to allow a very small degree of settlement to take place whilst the mortar is setting, before complete removal the following day. **3)** Again, in traditional formwork (shuttering), used to form in-situ reinforced concrete floors, *wooden- or telescopic metal props* were *eased* (slackened very slightly) the next day after the concrete had been poured, again to allow a very small degree of settlement to take place, before eventual removal.

Easiclean hinges: See *Figure 17* on page 42: These face-fixing sherardized hinges – used as an optional alternative to the more common storm-proof cranked hinges used on

Stormproof *casement windows* – produce a large gap on the hinge-side when opened, for cleaning the exterior glass surfaces.

Eaves: See *Figures 11 or 67* on pages 22 and 231: The *eaves* are the lowest edge of a roof slope, which usually overhangs the structure by between 200mm/8 in., up to about 450mm/18 in. – and where the rainwater drainage is placed via a system of guttering and RWPs (rainwater pipes).

Eaves' carriers/felt-support trays: These are a relatively modern addition to roofing, or re-roofing, and they give good support to the otherwise-vulnerable *eaves' edges* of the roofing-felt (or breathable-sheet membrane) placed over the pitched rafters. The carrier's firm, plastic outer-edges also provide better conveyance of any errant rainwater into the guttering.

Eaves' plate: A timber wall plate of a 100mm x 75mm or 100mm x 50mm sawn section, with (uncommonly) no wall-support below it, that spans over a window to carry the birdsmouthed feet of the roof-rafters above it.

Efflorescence-staining: This term refers to powdery white crystalloid salts that are released from newly-laid brickwork as it dries out. This white, patchy staining to otherwise-acceptable face-brickwork is unsightly and can cause concern, but it does gradually disappear over a period of time.

Electrical engineer's report: During a building-survey on a property, a Building Surveyor usually takes note of the condition and safety factor regarding electrical fixtures and fittings, any visible surface-wiring or cabling, consumer-units or fuse-boards and boxes, etc., and decides whether to recommend an electrical engineer's survey and report to be carried out.

Electrolytic corrosion: This refers to the corrosion of metals which can be caused by bringing two dissimilar metals in contact with each other. The water regulations require that no metal pipe or fitting shall be connected to other pipes or fittings of dis-similar metal, unless reliable measures are used to prevent damage caused by electrolytic corrosion.

Elliptical staircase: A geometrical stair formed within a stairwell that is elliptical in plan and spiral (or partly spiral) in elevation.

Enamel: A hard-gloss paint containing a high proportion of varnish with reduced pigment (colour). Because of this reduction, enamel paints require good, well-applied undercoats; at least two in addition to the priming coat.

Encaustic tiles: Such wall- and floor-tiles were made mostly by hand in the mid-19th century, by using a variety of coloured clays that permeated through the entire thickness of the tiles to form a variety of pre-designed patterns. Nowadays, encaustic-*effect* tiles are made in a variety of patterns and colours that effectively simulate the Victorian originals.

Enclosed staircase: This refers to a flight of stairs with a wall on each side, i.e., a stair fixed between two walls. To meet AD K1, of The Building Regulations, such a stairway requires a handrail to be fixed to at least one of the side walls if the stair is less than 1m – and two handrails (one on each side) if the stairway is wider than 1m. Note that enclosed stairs are sometimes referred to as *cottage stairs*.

Energy-performance certificate (EPC): This is a mandatory certificate required upon the sale or resale of a dwelling. The vendor can arrange for a trained inspector to carry out the mini-survey required to produce an EPC, but to expedite matters, it is usually the Estate Agent who initiates this via their own established contact. The energy-efficiency rating is based on such factors as whether a dwelling has cavity walls, sealed-unit double-glazing, insulated lofts and/or insulated walls and ceilings to loft/ attic rooms – and energy-efficient lighting, etc., and this rating influences the environmental impact (CO_2 rating) – but, whether it influences the prospective purchaser is another matter.

Engineered joists: *Figure 37*: Over the last five or six decades, a wide range of so-called *engineered joists* and *beams* have been developed and used to construct suspended floors in dwelling houses, etc., as opposed to the use of traditional solid-timber joists. The latter are still seen to be used, but engineered joists have many undeniable advantages. One of them being that they can be designed for *single-floors* over greater spans, without the need for intermediate load-bearing walls, or mid-span supporting cross-beams. Engineered joists nowadays range from timber I-shaped joists with varying section-sized

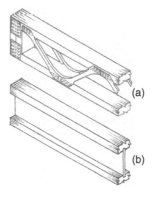

(a)

(b)

Figure 37 Two examples of modern, engineered floor-joists: (a) with metal webs and (b) with 'oriented strand board' (OSB) webs

timber top and bottom flanges, grooved to receive varying, central web-thicknesses of timber strand-board, similar to oriented strand board (OSB), to another system of joists that uses a similar arrangement of top and bottom stress-graded timber flanges, combined with a unique open-web design of zigzag-shaped galvanized-steel fixed to each side of the flanges. At least two advantages of this system are 1) The joists are relatively light in weight and 2) If required, concealed pipes, ducts and/or electrical wiring, etc., can be easily laid across the bottom flanges, through the zigzag web-systems.

Engineered-wood flooring: See **Laminate flooring**

Engineering bricks: These uniquely uniform-sized, dense building-bricks have a high crushing strength and – by virtue of their density – suffer very low water-absorption. For this reason, they are sometimes used to act as a damp-proof course (DPC).

English-bond brickwork: *Figure 38*:   An elevational pattern of *facing* bricks comprised of alternating, repetitive rows (*courses*) of *stretcher* bricks, in relation to alternating, repetitive rows (courses) of *header* bricks. The fact that bricks are theoretically 225mm/ 9 in. long, usually indicates that when header bricks are on display, such bonded walls are of *single-skin* construction (often referred to as being a *solid* wall). The exception is when – to give the impression of a solid wall – snapped *headers* (half-bricks) have been used.

English garden-wall bond: A one-brick-thick (225mm/9 in.) wall-bond, predominantly built of stretcher-bond brickwork on each side, but with a reinforcing course of header-bricks across its thickness at every rise of three or five courses. The main reason for using this bond is to achieve a *fair-face* finish on each side of the wall – which is easier to achieve with a predominance of stretcher bricks.

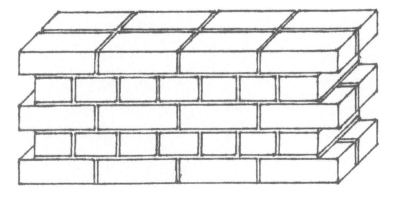

Figure 38 English-bond brickwork

Epoxy floors: This term refers to epoxy-resin coatings that provide a hard-wearing surface to certain floors, such as a concrete garage-floor.

Equilibrium moisture content: This expression refers to the desirable moisture-content of timbers or timber-components used in habitable buildings, when they neither gain nor lose moisture via being influenced by a balanced condition of humidity and temperature.

Ergonomics in kitchens: See **Work triangle**

Escape-stairs: See **Fire-escape stairs**

Escutcheon: With origins in French heraldry and ancestral shields, *escutcheon* in building terms refers to a small oval- or round-shaped metal plate that pivots over a keyhole (on the room-side of a door) to protect (shield) the room's privacy.

Espagnolette bolts: **1)** Traditional, surface-fixed bolts that are fixed to the inner face-side of one of the rebated meeting-stiles of double, glazed doors (French doors). The stapled 'bolt' occupies the full height of the rebated door and is operated by a central, levered handle. This, when turned, 'shoots' the bolt into the top and the bottom staples. **2)** Modern Espagnolette-type bolts are now widely used on the closing edges of uPVC and composite doors – and are much more sophisticated, with multi-locking side-edge bolts operated by an upward lift of the levered handle and then locked by the turn of a key.

Estimates: Legally, whether given to someone verbally or in writing, an 'estimate' for work to be carried out, is an imprecise price, which may cost *more* or *less*. And it does not usually refer to VAT. For comparison, see the definition for **Quotations**.

Expanded metal lathing (EML):   This traditional plastering material was often referred to as *wire mesh* and consisted of thinly-gauged (about 1mm/¹⁄32 in. thickness) sheet-steel, pierced with close lines of staggered cuts, then expanded mechanically to form an open network of 6 to 9mm/¼ in. to ⅜ in. diamond-shaped mesh. Then coated with bituminous paint, it was obtainable in rolls of 686mm/27 in. width and was fixed tautly across timber studs and/or joists as a substrate to fire-resisting plastered surfaces.

Expanded polystyrene: This low-cost polymer of styrene, insulating material, mostly used in sheet-form nowadays, is in varying thicknesses from 25mm/1 in. to 100mm/ 4 in.

Expansion joints: As covered in this book under **Thermal expansion and/or contraction**, where it is explained that all materials and forms of construction expand and contract via differences in temperature, causing thermal movement and likely damage, this heading now covers the use of a variety of expansion joints that are used to combat the anticipated damage that thermal movement can cause.

1) Large or medium-sized areas of concrete to be laid as driveways, or parking areas, etc., should be divided up into bays, so that each attached bay is separated by an expansion joint. The expansion-joint material can be of 13mm/½ in.-thickness bituminized fibreboard – which is cut to suit the concrete depth – and is usually self-supporting against the previously-shuttered edge of a recently-laid bay.

2) When the design-length of exterior masonry-walls are considered to be vulnerable to damage via excessive thermal expansion and contraction, *vertical expansion-joints* (wall-tied gaps of about 13mm/½ in.) are designed into the wall-length, shown on the contract-drawings and specified to be built into the walls. At an early stage after such walls have been built, the vertical gaps are filled (caulked) with coloured, exterior masonry sealant. The near-matched colour usually hides their existence to some extent – but they are far less disturbing visually, than the otherwise likely appearance of diagonal thermal-cracks emanating from the corners of window- and doorway-head openings in a wall.

3) Traditional-style floorboarded rooms, modern tongue-and-grooved floor-panelled rooms (usually of chipboard) and overlaid floating floors, all have to have expansion joints to combat thermal movement. In the first instance, the ends and sides of the floorboards laid and fixed directly on the timber floor-joists should be kept away from structural walls (or partitions) by approximately 12mm/½ in. This is partly to lessen the risk of picking up dampness from a wall, but mostly to lessen the risk of thermal expansion causing the boards to buckle upwards in their mid-area, after expanding and touching the walls.

4) In the second instance, overlaid, floating floors of engineered hardwood or plastic laminate, also require expansion joints at their perimeter edges, which either involves taking off and replacing the skirting boards to achieve a secreted expansion joint – or is done by taking a DIY approach, by placing the 12mm/½ in. thick cork expansion-strips *against* the skirting boards and covering them by fixing 19mm/¾ in. quadrant- or scotia-shaped mouldings to the bottom surfaces of the skirting-boards, after the floor is laid. However, if the skirting boards are at least 150mm/6 in. deep, it would look more professional to fix a second skirting member of, say, 75mm/3 in. depth to the lower faces, rather than the amateurish-looking quadrant- or scotia-moulding. And this would not be unprecedented in second-fixing carpentry; traditionally, *built-up skirting boards* (with two- or three-stepped tiers) were quite common in large properties. Although such *stepped tiers* were achieved via the skirting boards being fixed to a built-up substructure of horizontal and vertical *timber skirting-grounds* – as opposed to being built-up entirely on each other.

Expansion pipe: See **Expansion tank**

Expansion sleeves: This term refers to cylindrical cardboard, plastic, or metal sleeves – usually called *pipe sleeves*, that are built into walls, ceilings or floors, etc., to act as oversized tunnels for lesser-sized pipes – thereby allowing for the likelihood of expansion.

Expansion strips: See **Expansion joints**

Expansion tank: This refers to a small tank above an *indirect cylinder* (sometimes housed in the roof-space) which receives a swan-necked *expansion pipe* from the cylinder below, to allow for the possible over-expansion (and overflowing) of the water upon heating.

Exterior paint: See **Water-based paints used externally**

Exterior plywood: This innocently ambiguous term only means that the *glue* bonding the plies together is *moisture resistant*, not the wooden plies – although such plywood will of course resist more moisture than *interior-grade* plywood.

Extract vent/fan: A concealed electric-fan that draws moist- or foul-air from a room – usually a bathroom, shower room, WC or kitchen – and releases it to the exterior of the dwelling or building. Such extraction is compulsory nowadays in rooms containing a WC, if such a room does not have an opening window to the exterior air.

Extrados: In geometrical terminology, this term refers to the curved top surface of an arch.

Eyebrow windows: Traditionally, for aesthetic reasons, these windows – which are similar to dormer windows in most respects – have a so-called serpentine-shaped roof over them, that is thought to look very much like a *raised eyebrow* – which the higher cost of constructing such a roof might cause.

F

Face mould: See **Handrail wreaths**

Facing bricks: Bricks of better quality, made purposely with attractive shades of colour and/or textures, for use on so-called facework, i.e., the exterior faces of elevational walls, etc.

Fair-faced brickwork: Because solid, non-cavity walls of half-brick- or one-brick-thickness (perhaps being built as garden walls, for example) cannot usually be built with facework on both sides, one chosen side has to receive a *fair-faced* finish. This means that the mortar joints should be filled in neatly and/or smoothly, but the general appearance will not be to the same standard as facework. In some cases, when the fair-faced side is too uneven – usually when slightly differing lengths of bricks protrude in a one-brick thickness (225mm/9 in) garden-wall bond – the fair-faced side is not pointed, as such, but the filled mortar joints are ***bagged-over*** with a crumpled-up piece of sacking. This equates to a *fair-faced* finish.

False ceilings: See **Suspended ceilings and floors**

False headers: This refers to half-length facing-bricks (*bats*, with two good end-faces) that were (historically) sometimes used as full-length *headers*, to build the outer-leaves (skins) of cavity walls in a Flemish-bond pattern (that also uses stretcher bricks), to give the false impression that the outer walls were of a more solid, one-brick-thickness of 225mm/9 in. But, to my mind – providing that wall-ties were used – no structural damage was caused. And this practice only reflects an undeniable truth that there is greater solidity in 225mm/9 in.-thick walls compared to the relative frailty of 112mm/ 4½ in.-thick walls.

   However, confusingly, half-length facing-bricks (bats) were also used to give the impression of full-length header-bricks in walls meant to be of one-brick-thickness (225mm/9 in.) Flemish-bond construction. This was done to minimize (by 50%) the use of the more expensive face-bricks. In reality, such walls – with their false header-bricks (and no wall-ties) – were/are structurally dangerous, by virtue of being two unbonded and untied half-brick-thick walls contravening structural slenderness ratios.

Fanlight window: Usually a top-hung, outward-opening casement window, over a window or door above the transom.

Fan truss: *Figure 39*: This is a reference to a common configuration of factory-designed, pre-assembled timber trussed-rafters used in modern-day roofing.

Fascia board(s) See *Figures 66 andlor 67*:   Similar to *barge boards* (explained under that heading), but fixed horizontally to the ends (the plumb-cut faces) of the common rafters at the eaves of a roof, to seal the openings between the rafter-ends and to carry the bracketed guttering.

Fat edges: See **Curtaining**

Featheredge-boarded fence: See **Close-boarded fencing**

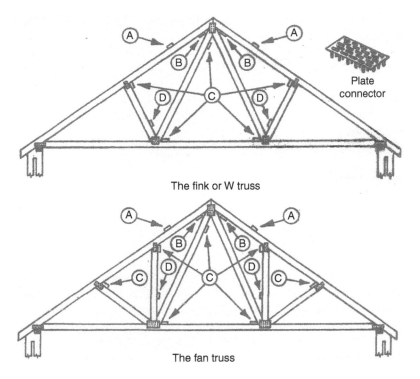

Figure 39 Fan-Fink- or W-trusses in modern roofing. (A) 25 × 100mm temporary, longitudinal timber bracing; (B) 22 × 97mm permanent diagonal bracing; (C) 22 × 97mm permanent longitudinal bracing; (D) 22 × 97mm permanent diagonal web chevron bracing.

Note that all truss joints are held together with galvanized steel plate connectors

FED: Front elevation/entrance door.

Fences: See Shared fences

Fender wall: This refers to a one-brick-thick (225mm/9 in.) low-rise wall, built-up from the oversite concrete for the purpose of containing the structure of the fireplace hearth, which must be not less than 500mm/19¾ in. away from the front of a chimney breast, nor less than 150mm/6 in. away from each side of the fireplace opening.

Fenestration: This word, in building-design language, means *the arrangement of windows and doors in the façades of buildings, to relieve their sheer, featureless appearance –* and in 2002, in response to certain changes in the Building Regulations, concerning a regulatory requirement for more stringent thermal insulation to *replacement* windows and *replacement* exterior doors, it (the word *fenestration*) was partly adopted by a new government quango to set up a regulatory body named **FENSA**, meaning **FEN**estration **S**elf-**A**ssessment.

FENSA: As mentioned above, this government body was set up to monitor compliance with the changes in Building Regulations concerning thermal insulation to *replacement* windows and *exterior* doors. And this monitoring appears to be done by licensing window- and door-installers/companies (via annually-paid licences) and supplying them with FENSA certificates (a signed copy of which should be given to a customer) to self-certificate their own work. It is not certain whether FENSA check up on this work occasionally. FENSA certificates provide an installer's customers with a 10-year guarantee. Note that if, upon the sale of a property, the solicitors learn (via their questionnaires) that any replacement exterior-doors and/or windows (*replaced after 2016*) are not covered by a FENSA certificate, the seller will be asked to pay for an indemnity insurance policy – to indemnify the buyer against possible non-compliance with the amended Building Regulations – even though such regulations are not enforceable retrospectively on the seller.

Ferrous metal: Containing iron and therefore susceptible to rust (iron oxide) corrosion.

Fibreboard: This generic term refers to a variety of boards or sheets (often spoken of as *sheet material*), that are comprised of compressed wood-fibres or wood-shavings, etc. They are obtainable in a range of different sizes; the most popular being 2.4m x 1.2m/ 8 ft. x 4 ft. – although larger sizes (and smaller sizes in DIY outlets) are available. Their thicknesses – with a few exceptions – usually range from 3mm/⅛ in. to 25mm/1 in. This range of sheet material includes oriented strand board (OSB), chipboard, medium density fibreboard (MDF) and hardboard, etc.

Fibreboard cladding: This is a composite material, using recycled products mixed with resin, as an alternative to timber or PVC cladding. It is extremely tough and durable, water-resistant and relatively maintenance-free. It can be obtained with a very realistic, indented wood-grain effect to a variety of colours.

Fibre saturation point: This term refers to the **moisture content (mc)** of seasoned timbers used in a building process, when – in certain situations – they exceed their optimum moisture content (by taking on extraneous moisture) and become more susceptible to timber diseases and decay. The moisture content is expressed as a percentage of its kiln-seasoned dry weight. A maximum of about 20% mc is usual for carpentry timbers and 12 to 15% mc for joinery. Note that in certain situations, such as when timber joists are exposed in an unvented cellar, the *wet rot/dry rot barrier* is reckoned to have been breached if their mc exceeds 20%.

Fibrous plaster: Traditionally, moulded ornate shapes for ceiling-cornices, etc. – as well as being formed in-situ on site with hand-made running-moulds – were also pre-made in workshops by casting in gelatine- or plaster-moulds. To give them the necessary strength to withstand being handled and transported to site, they were reinforced with coarse-woven canvas, wooden laths and wire netting, etc.

Fillet: **1)** A small, rectangular, timber cover mould; **2)** a small, triangular sand-and-cement weathering-seal, sometimes found on traditional lean-to roofs, when a side-abutment of the roof buts up to a brick-wall. Such fillets are generally regarded nowadays as being of a very inferior practice – having been used traditionally where soakers and stepped flashing (not to mention *cavity trays*) should have been used.

Finger joints: See **Comb-** or **Finger-joints**

Finger (push) plates: These come under the description of *door furniture* and were/are used to protect the face-sides of door-stiles (just above the door knobs or lever handles) from grubby, greasy finger-marks/staining. They were/are usually of copper, brass or aluminium on office doors and of ornate ceramics on those used historically in dwelling houses.

Finial: **1)** A short, wooden-post, usually with lathe-turned ornamental ends, traditionally fixed at the end of a ridge board, sandwiched between the plumb-cuts of the paired barge boards. If not maintained (repainted) regularly, such ornamentation is very prone to wet-rot decay; **2)** Alternatively, finials, in the form of clay-fired ornamental objects (such as eagles, dragons, etc.), can also be seen on a number of old-build properties, as an integrated part of the end ridge-tile above the apex of the barge boards.

Finished floor level (FFL): This is a very important reference on most building-site projects and must be strictly adhered to by all relevant trades, as the work progresses. It relates to the OBM (ordnance bench mark) datum levels already set up or being used on new-build sites, to enable different trades such as bricklayers, carpenters, drain-layers, etc., to measure *up* in relation to floor-levels, or *down* in relation to excavating and forming inspection chambers (manholes) and underground drainage systems.

Finishing coat: **1)** In wet-plastering techniques, this term refers to the approx. 3mm/⅛ in. thickness of fine-powdered finishing plaster, that is applied to the coarse-rendering coat of plaster, to achieve a smooth surface-finish; **2)** In painting and/or decorating, the term refers to the final coat of paint, or varnish, etc.

Fink or **W truss:** See *Figure 39* on page 101: This is a reference to one of the most common configurations of factory-designed and assembled timber trussed-rafters used in modern-day roofing.

Finlock-type guttering: *Figure 40*: This precast-concrete system of roof-guttering, which originated soon after WWII in the early 1950s, became discredited over time via leakage problems. They were apparently made by a few companies, to slightly different designs, but have acquired the generic name of *Finlock guttering* – such being the name of one of the manufacturers.

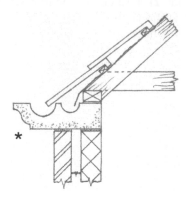

Figure 40 Finlock-type concrete gutter*

They can still be seen here-and-there serving a number of pitched roofs of post-war houses and bungalows, but such scrutiny does not confirm possible leakage problems. They vary slightly in their precise design-features, but a full-length gutter would be comprised of a number of butt-jointed precast units of about 600mm/2 ft. length x 400mm/16 in. width x 100mm/4 in. height. Within their width, two concaved channels run side-by-side throughout their length. One being the gutter-channel, the other (which is concealed beneath the tiled eaves' projection), is believed to be an inner gutter to serve the roofing-felt overhanging into it. One of the end units has a 75mm/3 in. diameter outlet-hole in its front channel to house the rainwater-downpipe connection. The shallow-depth front face of the units (housing the gutter channel) has an ogee- or cyma-reversa shape. The units are bedded on top of 280mm/11 in. cavity walls (with the gutter-channel side projecting by one-third) and are apparently butt-jointed with a bitumastic-compound – which is their Achilles' heel. Over a period of time, the natural expansion and contraction via ubiquitous thermal movement of both the guttering and/or the wall-bearing brickwork, cause the joints to leak. And this leakage is known to cause descending-damp problems to the interior walls.

Although periodic bituminizing of the gutter's concaved channel is advisable, it is likely that any particular Finlock system that has an inner, concealed channel (as described above), might mean that the leakage problem can only be half solved. Also, any thoughts of removing and replacing a Finlock-type system with, for example, uPVC guttering are also problematic – because the roof appears to take its bearing on the inner, top surface of the concrete guttering.

This, therefore, means that if the Finlock guttering needs to be replaced with more common plastic guttering, the non-projecting eaves of the roof (initially the common rafters) need to be extended to create eaves' projections.

Fireback: The structural wall behind the fireplace-opening.

Firebricks: Such bricks are made from fireclay, or refractory clay that has been quarried from coalmining areas and contains a high proportion of silica and a small proportion of lime and iron. These clays are very resistant to high temperatures.

Fire cement: Made from refractory cement, this ready-mixed compound – that resists cracking when exposed to heat – is used to joint spigot-and-socket type flue pipes attached to boilers, etc.

Firedogs: A pair of H-shaped ironwork-stands used for supporting burning logs in an open fireplace.

Fire-escape routes (including **windows**)**:** The Building Regulations, Approved Document B1 (with amendments) covers *Fire Safety in Dwelling Houses*, where there is reference to an upgrading of smoke alarms in *new* dwelling houses – and, in Section B2, there is reference to providing *Means of escape* via emergency egress from ground-level and upper-floor-level windows (or doors).

These provisions refer to *Escape from the ground-storey* – and *escape from upper floors not-more-than 4.5m above ground level*. They also refer to *Escape from upper floors more than 4.5m above ground level*. But, additionally, these other categories of dwellings must also provide a fire-protected staircase and other criteria, such as *having no inner rooms above ground-level* and/or providing an alternative escape route or a sprinkler system, depending on whether there is one floor above 4.5m, or more than one floor.

Basically, new windows that are provided for emergency egress/means of escape, should have an unobstructed openable area of at least 0.33 m² and be at least 450mm high x 450mm wide. The bottom window-ledge (or sill) of the escape-opening should be not-more-than 1.1m/3 ft 7¼ in. above the dwelling's floor level.

The window or door – based on a reference to one's judgement and other criteria in B2's General Provisions in Section 2.8   should enable a person escaping from fire to reach a safe place. Note also that there is a reference to *Approved Document K, Protection from falling*, when *guarding* might be relevant.

Note that it also states under section 2.8, that locks (with or without removable keys) and casement stays may be fitted to egress windows, subject to the stay being fitted with a release catch, which may be child resistant.

And *Note 3* of paragraph 5 states that: 'Windows should be designed such that they will remain in the open position without needing to be held by a person making their escape.'

Fire-escape stairs: Such stairs are required by Building Regulations in a number of buildings *other than* private dwelling houses. But if a private dwelling house has been converted into flats, it becomes an HMO (a House in Multiple Occupation) and must either have a flight of stairs outside of the building, or the inside staircase must meet (or be upgraded to) the regulatory requirements of a so-called *protected shaft*. This might mainly involve changing any ordinary doors serving the stairwell, to fire-resisting doors – and changing any glazed panels, borrowed lights or glazed door-apertures with fire-resisting safety glass

Fire-grate: An iron-barred grid arrangement with four short legs, that supports kindling wood and solid fuel in an open fireplace.

Fireplace: The indented recess in a chimney-breast, where the fire-grate is placed.

Fireplace hearth: See **Hearth/hearthstone**

Fire protection of structural-steel framework: The bombing of London and other UK areas in WWII in the 1940s, causing total destruction of many steel-framed buildings in the cities, brought a realization that bare, unclad steel framework (as existed then) in mostly large buildings, behaves badly when such buildings are on fire. The morning after the nightly air-raids, large buildings were found to be in total ruins via their twisted and buckled steel framework. Such heat-distortion had caused the buildings' floors and walls, etc., to collapse. Since that period of realization, steel girders (RSJs), beams and stanchions, etc., used in *any* building construction, are usually encased with a protective covering of a variety of materials, including concrete and fire-resisting-grade plaster-board, fire-retardant paint, etc.

Fire-resisting (FR) doors and frames: Nowadays, these are referred to as *fire-resisting doorsets*, with a quoted stability/integrity rating. This rating is expressed in minutes, such as 30/30 or 60/60, meaning 30 (or 60) minutes' stability, 30 (or 60) minutes' integrity. Stability refers to the point of collapse, when the doorset becomes ineffective as a barrier to the spread of fire. And integrity refers to holes or gaps concealed in the construction when cold – or to cracks and fissures that developed under test.

Fire-resisting glazing: New or replacement (clear-or-obscured) glass that is required by Building Standards (BS) to be *fire-resistant*, should conform to BS EN 12150 (for toughened glass), or BS EN 14449 (for laminated glass). Traditional-style GWP (Georgian Wired, polished Plate, or obscured safety glass of 6mm/¼ in. thickness), has a 30-minute fire rating and provides a degree of safety via its embedded wire-grid (which prevents dangerous shards of glass from scattering if breakage should occur).

Fire-retardant paint: Such paint is used to protect unclad steel components, such as steel beams (RSJs) and steel stanchions (H-shaped columns), etc. Usually, these unclad members can be seen in industrial and/or commercial steel-framed buildings, such as supermarkets. The fire-retardant paint inhibits/retards the spread of flames by releasing a flame-dampening gas once the painted surfaces reach a certain temperature. Fire-retardant paints must conform to BS 476/EN 13501.

Fire-stops: This refers to built-in fire barriers, comprised of incombustible material, usually in unseen, concealed places, between otherwise separated fire-zones.

Fire-surround: A plain, or ornamental structure, usually with a mantelshelf, built or fitted against the face of a chimney-breast, at the sides and top of an indented, open fireplace.

Firring-pieces/blocks: These terms refer to parallel or, more commonly, wedge-shaped (tapered) pieces or lengths of softwood timber, nailed onto the top edges of joists, etc., to achieve a level, a sloping, or a higher surface when boarded or covered in sheet material. A typical, common use of firring-pieces is to create a *fall* (a slope) on a 'flat' roof.

First-, second- and third-fixings: These terms refer to the different stages of trade-operations to be carried out by carpenters, plumbers and electricians during the construction process. For *carpenters*, first-fixings include modern- and/or traditional roofing, fitting and fixing modern and/or traditional floor joists, etc., *before* plastering or drylining of walls and ceilings takes place; second-fixings include fitting staircases, hanging doors, fixing skirtings and architraves, etc., *after* dry-lined plastering has taken place and the shell of the building is watertight; and third-fixings include fitting and fixing ironmongery such as locks- and/or latch-lever-furniture to doors after priming and undercoat painting has taken place. For *plumbers*, first-fixings include fitting and fixing the large diameter soil-and-vent pipe(s) in the building, ready for an immediate watertight exit though the roof, once it is clad; fixing lead-flashings where required; fitting a stopcock to the mains water supply at ground-floor level – and fitting and fixing the gutters and rainwater pipes as soon as is possible after roof- and eaves-cladding; second-fixings include fitting and fixing WCs, baths, showers, basins, sinks, boilers, etc. For *electricians*, first-fixings can include the fixing of conduit and terminal boxes, plastic trunking (or laying ring-main cables and/or lighting cables), prior to the second-fixing operation of wiring and the third-fixing operation of wiring-up and fixing the light- and power-fittings, etc.

Fixed light: A window that does not open.

Flagstone: A flat slab of natural sandstone, of varying thicknesses and superficial sizes, but usually not less than 25mm/1 in. thick. Traditionally used for street-paving and as manhole (inspection-chamber) covers. The latter would be oblong-shaped and at least 38mm/1½ in. thick.

Flame spread: This refers to the testing and approval of sheet-materials used in the UK for lining walls and ceilings in accordance with BS (British Standards) criteria. Materials are, apparently, tested under intense radiant heat and a gas flame, to determine how much time elapses before their exposed surface ignites. Their rate of flame spread is then rated as being: *very low, low, medium* or *rapid*.

Flammable: For some years now (because it caused confusion) the word *flammable* has replaced the word *inflammable* – although both mean the same thing: *Something which is easily set on fire*. **Non-flammable**, therefore, means: *Something which is not easily set on fire*.

Flashing: See *Figure 30*: A protective 'apron' that covers various roof junctions where brickwork pierces through the roof, such as at chimneys and low-rise fire-break party walls, etc. Where **plain tiles** and/or **slates** are involved, right-angle-shaped **soakers** must also be used – under the flashing, between the tiles or slates. However, **interlocking tiles** do not require soakers – they only require a wide, **stepped** *cover*-**flashing**; the extra width of stepped flashing allows the un-stepped edge to be dressed over the first raised- or sunken-feature on the interlocking tiles. The most common materials used for flashings and soakers are sheet-lead, copper, zinc and mineral felt; lead and copper are best for longevity.

Flat-conversions: Over quite a long period of time, single-dwelling houses (large and small) have been converted into flats and/or self-contained rooms with shared facilities. Initially, this was done without the need for stringent local-authority Building Control involvement, although such conversions had to meet minimal fire regulations and be re-rated for separate Council Tax charges. However, since then, the local authorities, wielding ever-increasing legislation, have been rightly battling with owners/landlords to upgrade these now-named HMOs (Houses in Multiple Occupation). Such upgrading was required regarding thermal- and sound-insulation, fire-alarm systems, protection against fire and fire-escape routes via windows and protected shafts (stairways), etc. Although, prior to the amended Building Regulations in 1991, and the introduction of HMO classification in 2004, many such properties were well below a good standard. Even so, it should be realized that sound-insulation (an extremely important issue) of the floors between converted flats has usually been upgraded to a far better standard in flats converted *after* 1991.

Flat roofs (traditional and modern):   See **Cold-deck flat roofs**; or **Hybrid flat roofs**; or **Warm-deck flat roofs**, written herein separately.

Flats: See **Flat-conversions** or **Purpose-built flats/apartments**

Flat-topped arches: Such traditional-style arches contain radiating voussoirs, but the tapered bricks do not conform to the usual segmental top (extrados) – they are further shaped to achieve a flat top – and their underside (intrados) that may appear to be straight, has in fact, a very slight rise of 3mm/⅛ in. to every 300mm/12 in. run, i.e., if the window- or door-opening is 900mm/36 in. wide, the total, segmental rise = 9mm/⅜ in. The reason for the slight rise, is/was because flat-bottomed arches were thought to give an illusion of sagging.

Flaunching: *Figure 41:*   In simple terms, *flaunching* is a thick, cross-segmental shaped layer of sand-and-cement mortar, which is laid and shaped (trowelled) around the bases of chimney pots, partly for the purpose of securing them onto the top of a chimney-stack's flued brickwork and partly to create a weather-sloped surface around the pots. These convex-shaped slopes, that inhibit rain from settling, should emanate smoothly from the four top-side edges of the stack's brickwork and should visually take on the appearance of an imaginary model-roof with a double, hip-ended domed-roof

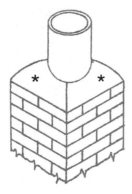

Figure 41 *Flaunching to a one-pot chimney stack

appearance, with pots protruding through the rounded apex area. Stiff flaunching-mortar should be used, comprised of **sharp** sand (not commonly-used **soft** sand) and cement, to a ratio mix of 3:1.

It must be understood that flaunching is in an extremely exposed part of a building and, over a period of time, it is extremely likely to develop mid-area surface cracks caused by thermal movement and aggravated by ingress of rainwater. And yet, because of this area of the roof's obscurity (and inaccessibility), such cracks – that lead to *descending damp* into the interior of a building – remain out-of-mind and undetected. However, it is usually possible for surveyors and/or house-owners to check the chimney-breast's brickwork from within the roof-void (loft). If descending damp is occurring, it is likely to look wet, or feel wet. It could, of course, be tested with a relatively cheap moisture meter. Any positive readings will obviously suggest that the flaunching and the chimney's mortar-joints' pointing, etc. – need to be checked-out above by a reliable roofer.

Flemish-bond brickwork: *Figure 42*: This refers to an elevational pattern of *facing bricks* comprised of a repetitive *header, stretcher, header, stretcher* formation on each row, whereby all headers are centralized over the stretchers – and the bond is terminated at the quoin with either ¾-sized stretcher bricks or ¼-sized *closer* bricks. As mentioned with English-bond, the fact that header bricks are on display, *usually* indicates that such bonded walls are of *single-skin*, solid-wall construction of at least 225mm/9 in. thickness – but, this is not always the case. Historically, there is evidence of *snapped* headers having been used to give the impression of a more solid wall.

Fletton bricks: These popular bricks, made of the clay from an area near the village of Fletton, Peterborough, are normally used as *common* bricks, i.e., for backing-up the face-sides of walls built with *facing* bricks, when such walls are more than a *half-brick* thickness and require additional bricks on their back-sides. Such thicker walls may be of so-called *one-brick* thickness, *one-and-a-half-brick* thickness, etc. The bricks' secondary uses include any structural, unseen walls, such as work below DPC level,

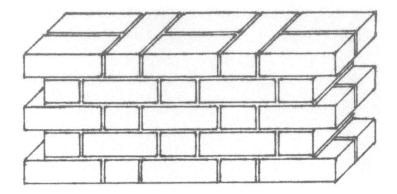

Figure 42 Flemish-bond brickwork

dividing-party-walls, designated to be plastered or dry-lined. However, it must be mentioned that, nowadays, such walls are usually built with high-density blockwork.

Flight: The part of a stair or ramp between landings that has a continuous series of steps or a continuous slope/ramp.

Flitch beam (or **flitched beam**):   This refers to a traditional built-up beam comprised of two on-edge wooden joists, reinforced with a relatively-thin, joist-depth steel *flitch plate* sandwiched between them. These three components are bolted together through predrilled centreline holes in the 6mm/¼ in.-thick steel-plate, spaced at 600mm/24 in. centres.

Floating floors: Such floors, comprised of tongue-and-grooved (T&G) flooring panels, edge-glued and laid in a staggered, stretcher-bond pattern, are technically described as having an advantage of *continuous support*. This is because they are laid on a flat surface (usually of rigid insulation sheets), rather than on the top, narrow edges of floor joists – and another advantage is referred to as having *discontinuous construction*. The former advantage is quite obvious, but the latter may not be, as it refers to the subfloor of insulation that separates the top layer of floating floor from the structural floor below. Such discontinuous construction is essential where sound insulation is required in addition to thermal requirements. As an example of a floating floor, laid on a damp-proofed, concrete oversite, it may consist of manufactured tongue-and-grooved (T&G) flooring panels, either of chipboard (preferably a moisture-resistant grade) or plywood, laid in a staggered, stretcher-bond pattern, floating on close-butted insulation sheets/slabs of 100mm/4 in. thickness, with a polythene vapour check laid between them. Note that floating floors are held down by their own weight and the perimeter skirting.

Float, render and set: This technical terminology refers to wall-surfaces that require (or are specified to receive) *two* rendered undercoats with a coarse material (such as a mixture of sand, lime and cement) and a *setting* of fine *finishing* plaster. Such a three-coat build-up is sometimes specified to achieve a flatter, less uneven finished surface.

Floor boards: In old properties, depending on their actual age, square-edged floorboards can be found with various widths ranging from ex. 225mm/9 in. to ex.150mm/ 6 in. (the wider the width, the older the age of a property is likely to be). Their thicknesses usually varied between ex. 25mm/1 in. to ex. 20mm/¾ in. They were usually nailed to the floor joists with so-called *black-iron floor brads*. But if the boards are really old – say, of the 15th century – and are fixed with wrought iron nails of irregular shapes and clasped-heads, such nails might be hand-made (hand-forged) – and the floor boards might be of oak, with their widths up to 300mm/12 in.

Square-edged floorboards, with the lesser width of ex.150mm/6 in. and a thickness of ex. 25mm/1 in., were in use well into the 20th century and were usually fixed with so-called *cut floor brads*. After this period, tongue-and-groove-edged floorboards of ex. 150mm/6 in. width took over (because of their obvious advantage of providing a draught-resistant floor at ground-floor levels, where air-vents created draughts). However, this board-width eventually gave way to a lesser width tongue-and-grooved board of ex. 100mm/4 in. to combat the amount of surface-joint shrinkage and cupping suffered by wider boards. Such boards are still used nowadays as a subfloor, if they can be afforded. And if not fixed by a modern nailing gun, they should be fixed with so-called lost-head wire nails and driven slightly below the surface with a nail-punch.

Floor brads: These thin, tapered *cut* nails, in black iron, were used extensively for fixing square-edged and T & G floorboarding – and held the boards down well – but, nowadays, either lost-head wire nails, in galvanized, sherardized or bright steel, are used. But these do not hold the boards down as well. Extremely thin, narrow, bright steel nails, fired from a nail gun, are also used extensively for floorboard-fixing. They seem to hold the boards down quite well and do not require to be punched-in.

Floor insulation: The Building Regulations' upgraded Part L1, concerning the conservation of fuel and power, is, of course, not meant to be applied retrospectively to existing dwellings. But – with regard to floor insulation – improvements could be made. For example, traditional ground-floors with suspended joists on sleeper walls could be upgraded by having 100mm-thick Crown Wool packed between the similar-depth joists, laid on support-netting turned up at the sides and fixed to the faces of the joists. Of course, it would be a tedious task removing and – if undamaged – replacing the floorboards or sheet-decking. Alternatively, of course, the existing ground floor (whether suspended or of oversite-concrete, topped with a sand-and-cement screed) could be lined with a vapour/moisture barrier, before being overlaid with a thin (3 to 5mm thick) foam, insulating underlay, prior to laying a floating floor of plastic- or hardwood-laminate. The insulation underlay could be thicker, of course, but this might mean a greater reduction from the underside of any affected doors.

Floor-joist restraint straps: *Figure 43*:   Modern construction methods, involving lighter-weight material in roofs and walls, have led to the need for anchoring straps,

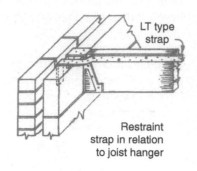

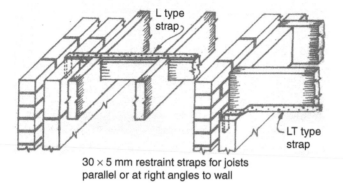

30 × 5 mm restraint straps for joists
parallel or at right angles to wall

Figure 43 Floor-joist restraint straps

referred to in the Building Regulations' Approved Document A, to restrict the possible movement of roofs and walls likely to be affected by wind pressure. Such straps are made from galvanized mild-steel strip, of 5mm thickness for horizontal restraint and 2.5mm thickness for vertical restraint. The straps are 30mm wide and up to 1.6m/5ft. 3in. long. Fixing holes are pre-punched along the length at 15mm/⅝ in. centres, to facilitate optional fixing points.

Floor joists: *Figure 44*:   The traditional floor joists outlined here are still unchanged in their usage, although a few modifications have taken place over time, such as pre-sale structural-grading of the timber, regularization of sectional-sizes and pressure-treatment with preservatives. They are still much in use nowadays, even though there are now more modern, diverse forms of floor-joists, covered herein under **Modern, designed floor-joists**. The joists referred to here are obtainable in various metric lengths, depths and thicknesses. These last two sectional sizes are dependent upon either **1)** the span across a room from wall-bearing to wall-bearing, or **2)** the span between any intermediary supports, such as **3a)** an RSJ (rolled steel joist) or RSJs placed across the shortest span, beneath the timber joists running at right-angles across them, over the longest span, (making such a construction technically a *double floor*), or **3b)** at ground-floor level, where traditionally, half-brick-thick, honeycombed sleeper walls were placed across the shortest span, beneath the timber joists running at right-angles

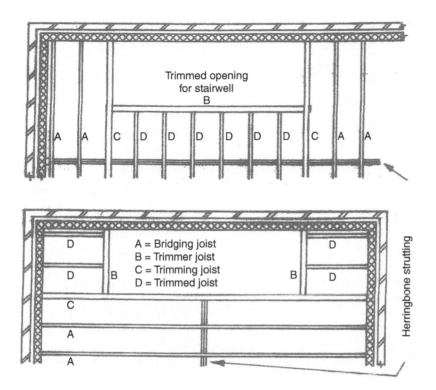

Figure 44 Part-plan views of alternative joist-arrangements around trimmed stairwell openings in a suspended floor

across them, the depth of joists depended upon the optional positioning of the intermediate sleeper walls.

Because of this advantage, it was quite common practice traditionally to always use 100mm x 50mm/4 in. x 2 in. sawn joists at ground-floor level, by building the wall-plate-bearing sleeper walls at no-more-than 1.6m/5 ft. 3 in. centres.

The different technical names of the timber joists listed below, is in the order of their structural-positions in a hypothetical floor-arrangement, to suit the formation of a suspended floor with a stairwell-opening against the mid-area of a flank wall. Note that two of the joists named here, the **trimmer** and the **trimming joists**, are 63mm/2½ in. wide, as opposed to the **bridging-** (or **common-joists**) and the **trimmed joists**, that are only 50mm/2 in. wide.

1) If the site drawings indicate that the floor-joist arrangement is to run parallel to the flank wall, then the first joist to be positioned is the trimming joist, bearing on the walls at each end – or fitted into joist-hangers – and set up away from (and parallel to) the flank wall by the required width of the stairwell. Next, it would be prudent to create a working platform, by setting up the bridging joists, parallel to the trimming joist, at their required spacings, to infill the bulk of the open area. Then, at each end of the stairwell, there is likely to be an area of floor required, so two trimmer joists should now be positioned, one at each end, at right-angles to the trimming joist. These small, open areas should then be infilled with trimmed joists, parallel to the trimming joist.

2) Alternatively, if the site drawings indicate that the floor-joist arrangement is to run at right-angles to the flank wall, then two trimming joists must first be set up at right-angles to the flank wall, with the stairwell length between them. Then, one long trimmer joist, bearing on the trimming joists at each end, is set up to the required width of the stairwell. Again, to create a working platform, a greater number of trimmed joists this time, should be trimmed into the trimmer joist. Finally, on each side of the long trimming joists, there will be a number of *bridging joists* needed to complete the skeletal framework.

Floor-numbering: See **Numbering and naming of floor levels in the UK and the USA**

Floor screed: Ideally, this should be a perfectly level, flat and smoothly-trowelled layer of *stiff* (semi-dry) sand-and-cement mortar, laid on a concrete substrate incorporating a DPM (damp-proof membrane). It is usually mixed to a 3:1 ratio of sharp sand and Portland cement. Its thickness should not be less than 40mm/1½ in., nor more than 65mm/2½ in. It must be allowed to dry completely before any finishing floor-material is laid or bonded to it.

Floor-spring hinge: This traditional, almost-unseen hinge was manufactured for doors that swing (pivot) through 180^0. It is comprised of an open-fronted metal shoe, connected to a metal floorplate via a pivoting pin, which enters the plate and connects to a sprung-hinge in a floor-concealed metal box. Once the door's frame has been set up, the floor-spring-boxed hinge is very carefully/precisely set up at the base of the doorjamb, ready to be permanently fixed by the eventual sand-and-cement floor screed. At the second-fixing stage of door-hanging, a separate projecting pin and a pin-holed receiving-plate are housed into the doorframe-head and the topside of the door, before the door is finally fitted into the receiving-shoe and side-screwed to the door.

Floor strutting: Solid strutting in between floor joists is used to give additional strength by interconnection between joists. It removes the unwanted individuality of each joist and effects equal distribution of any load to prevent joists bending sideways. Struts should be used where spans exceed 50 times the joist thickness. Therefore, with 50mm thick joists, a single row of central struts should be used when the span exceeds 2.5m., and two rows are required for spans over 5m. and up to 7.5m.

Note that the practice of strutting a floor with solid struts (equal to the joist-size) was frowned upon traditionally as adding unnecessary weight and creating an inflexible floor, but nowadays, there are at least two exceptions where it is recommended: (1) Section 5 of New Build Policy's Technical Manuals, recommends that *solid strutting should be used instead of herringbone strutting where the distance between joists is greater than three times the depth of the joists*; and (2) the TRADA (Timber Research and Development Association) document *Span tables for solid timber members in floors, ceilings and roofs for dwellings*, recommends that solid strutting should be used for certain long-span domestic floors.

Floor tiles: Available in a variety of materials and sizes, they range from: **1)** plain ceramic tiles; **2)** patterned ceramic tiles; **3)** porcelain tiles; **4)** quarry tiles; **5)** quarried slate; and **6)** quarried stone. Note that **1)** and **2)** above (which are also available as wall tiles) must be at least 9mm/⅜ in. thick to be used as floor tiles – especially when being laid on suspended, boarded-floors, fixed to floor-joists – where, in fact, *any* tiles are extremely likely to crack. To avoid this, tile manufacturers usually recommend that such floors should be overlaid with 18mm/¾ in.-thick plywood, with butt-joints recommended to be screwed at 150mm/6 in. centres. Ideally, therefore, ceramic and porcelain floor-tiles, are more suitable for laying on solid, sand-and-cement screeded floors.

Other materials, such as vinyl, cork and carpet are available as tiles – although carpet- and unsealed-cork tiles seem unhygienic in lavatories and bathrooms with WCs. Note that thermoplastic tiles, widely used about three to four decades ago, are now known to contain chrysotile asbestos fibres – so removing them can be hazardous.

With the exception of a sand-and-cement bedding-mortar that would be required for laying exterior quarry tiles, quarried slate and quarried stone (named above), most other tiles are bedded on premixed, tubbed tile-adhesive.

Flue: A vertical emission shaft, measuring roughly 225mm x 225mm/9 in. x 9 in. across its width and depth, that rises up via snake-like bends, from an open fire in a fireplace-opening formed in a traditional brick-built chimney breast. It is roughly rendered with brick-mortar.

Flue-gathering/throating: This refers to the inverted V-shaped sides of a fireplace opening, as it leaves the fireplace and diminishes in width to meet the 225mm x 225mm/9 in. x 9 in. flue opening.

Flue-lining pipes: Fireclay pipes of 225mm/9 in. diameter, that were jointed together as they were built into traditional brick-built chimney breasts. This practice superseded the previous and tedious time-consuming bricklayer's task of rendering (pargetting/parging) the brick-flues' interior surfaces as the chimney's flues were being built.

Flush-joint pointing: See **Pointing**

Fluting: See *Figure 5* on page 9: This decorative feature, consisting of a series of semi-circular grooves, can be found around the shaft of traditional, Ionic columns and on the face-side of rectangular-sectioned architraves, or other traditional moulded work. When used on architraves, plinth blocks were used at the base and ornate cornice blocks replaced the mitres at the head.

Flying buttress: This term refers to a steeply-pitched (and sometimes tiered) support to a building, that projects from the face of an external wall and is either built of brickwork or stonework, in keeping with the main structure's material. As opposed to common buttresses, *flying* buttresses are not attached to the supported building's lower façade. Part of its structure allows for a passageway opening at ground-floor level, against the building – usually with a quadrant-shaped arch at its head, bringing the buttress back to the supported building.

Occasionally, a flying buttress is actually accommodating chimney-flues from an adjacent low-rise building – the extremely tall chimney thereby gaining its required height above the adjacent taller building, in the form of an imitation flying buttress.

Flying chimney-breasts: This expression refers to a modern-day practice by DIYers or ill-informed builders, of removing chimney-breast projections from one or more rooms, leaving a remaining upper portion of the breast and/or chimney stack unsupported. Very occasionally, in the roof-void (loft space), it can be seen that an amateurish attempt has been made to hold up the remaining breast with pieces of wood – or an RSJ beam can be seen to span across the house, supporting the base of the curtailed chimney breast – and taking its bearings on the tops of the exterior load-bearing walls.

Another structural concern, not seemingly addressed by chimney-breast removers, is that when a substantial pier-like wall (such as a chimney-breast) is to be removed, will its removal effectively lengthen the wall and thereby violate the slenderness-ratio rule?

Flying freehold: This refers to freehold properties that are linked together or conjoined in such a way, as to make it legally difficult for one or the other of the freeholders to carry out any maintenance or repairs, without the other party's consent. An example of flying freeholds is when the owner of a large dwelling house, historically converted it into separate flats and arranged for each of the flats to have a share of the freehold. This is particularly onerous if the HMO (House in Multiple Occupation) was converted prior to 1991, when the Building Regulation's requirements of separation were far less stringent.

Flying shores: *Figure 45*:   These traditional forms of *shoring*, which also include *raking shores* and *dead shores* (covered separately in this book), are so-designed to give

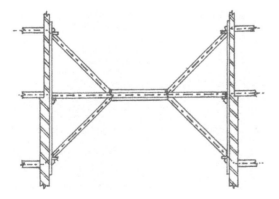

Figure 45 A basic elevation of a traditional 'flying shore' giving temporary support between terraced buildings

temporary support to one or two buildings whilst structural repairs or demolition tasks are carried out. Flying shores are usually required when a mid-terrace building is to be demolished, structurally repaired or altered. Basically, one or more flying shores (three being usual) are built within the structure of the subject building, to give support to the party walls of the other buildings on each side. This support is usually required until the subject building is either demolished, repaired or structurally altered. Such work may also involve *underpinning* of the party walls. Traditional flying shores consisted of a mid-area *horizontal shore*, resting on housed-*cleats*, with one wedged-end against one of the two vertical wall plates, four raking *struts* that formed a 45° X (cross) appearance and were connected to the wall plates via so-called hardwood-*needles* (that were cut into the walls) and housed-cleats at their wall-plate connections. Where the ends of the struts were seated on the top- and under-sides of the horizontal shore, their plumb-cut faces were wedged to two *straining-pieces* fixed thereto.

Note that the basic principles of flying shores might nowadays be incorporated in the form of tubular-scaffolding arrangements, using a minimum of timber components – but an enormous amount of tubular scaffolding and diagonal tubular bracing.

Folding shutters (also referred to as **boxed shutters**):   These traditional terms refer to two pairs of panelled shutters, each pair hinged to each other and each pair also side-hinged to the opposite, splayed side-reveals of a boxframe sash-window. The two pairs of folding shutters can either be opened up across the window – meeting each other on rebated edges – or each pair can be folded up and folded away into boxed recesses preformed in the splayed window-reveals. Note that in the opened-up position across the window, an end-pivoted, flat wrought-iron bar was hung in the mid-area of the hinged meeting-stiles on the right-hand side, ready to be pivoted from its dormant, vertical position to rest across and into two minimally-protruding U-shaped latch-plates on the left-hand side shutters. Folding/boxed shutters are still commonly found in a number of Regency and early Victorian properties.

Footings: *Figure 46*:   Although this term is still sometimes used loosely in the industry to refer to modern-day concrete foundations, it correctly refers to three, four or more,

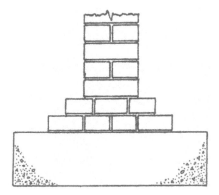

Figure 46 Brick footings and concrete strip-foundation (traditional)

stepped courses of brickwork that were historically built directly onto the ground (or onto hogging, without the use of concrete), to act as a foundation. When concrete started to be used for strip-foundations, brick-footings were still used (as illustrated) for a period of time. The brick-footings – being necessarily wider than the walls, to spread their load – stepped down on each side by 45° to their widest base-width. So, a brick-rise of 75mm equalled a 75mm step x the number of steps on each side of a wall. Therefore, a one-brick-thick wall of 225mm thickness, with three footing courses (3 x 75mm step-projections each side), required a first-footing width of 675mm, a second-footing width 525mm, and a final footing-width of 375mm, in readiness for the 225mm wall.

Formwork: This is a formal reference to preformed, temporary wooden structures, that are built to a required shape, to hold fluid concrete in place until a chemical setting-action takes place and the concrete hardens. Note that the term *formwork* is also referred to as *shuttering*.

Foul-water sewage: This term is often referred to as *soil*-water *drain* or sewage, but both mean fouled/soiled water from WCs, as opposed to *unclean* water from wash basins and sinks. The essential difference is that fouled/soiled water from WCs must be vented via a combined soil and vent pipe (SVP), whereas unclean water (via waste pipes) can discharge into an open, trapped gully, a hopper head or into the side of a 110mm/4¼ in. diameter SVP, via a side-bossed entry. The other important point regarding SVPs is that their open-tops should be at least 900mm/35½ in. above any opening windows at ground-floor level and at least 3m/9ft 10⅛ in. away from any opening windows at first-floor level. The SVP's open top should also be fitted with a vent terminal – to inhibit birds from nesting.

Foundation pads: *Figure 47*:   These *cast in-situ*, concrete *bearing-pads* are sometimes used to support individual loads, such as bolted-down steel-stanchions supporting a fore-court canopy; timber-posts, of a 100 x 100mm/4 x 4 in. section (seated in metal-pronged

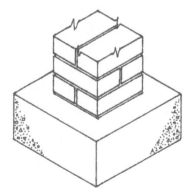

Figure 47 Concrete foundation pad supporting a brick pier

post-supports), and used to support the outer edges of a timber-framed balcony, etc. Traditionally, pad-foundations were also used to support low-rise 225 x 225mm/9 x 9 in. brick-built piers, used below timber-joisted ground floors – to support timber cross-beams that carried the 100mm/4 in. shallow-depth joists, in the (planned) absence of honeycombed sleeper walls.

Foundations: See **Strip foundations** and **Trench-fill foundations**

Foundation stone: Although named a *foundation stone*, it is usually – and sensibly – found well above actual foundation level, at least 300mm/12 in. above the paving level; it is usually positioned at the quoins (corners) of public buildings like Town Halls, etc. And such stones bear an engraved inscription on their face sides, the date of being laid, and the name of a dignitary who symbolically laid the stone (by tapping it with a trowel, or by laying the first trowelful of bedding mortar before the stone was professionally bedded and laid).

Framed, ledged, braced and *battened* **door(s):**   Note that the italicized term (*battened*) is to inform the reader that this historic reference is not used nowadays; if it was used, it would be replaced by the term *matchboarded*. But, in fact, these doors are now commonly referred to as FL&B doors *(Framed, ledged and braced) doors*. They consist of a three-piece *solid* (full thickness) frame, namely a top rail and two stiles (which are mortise-and-tenoned together), two reduced-thickness ledger rails, with bare-faced tenons into the stiles (their reduced thickness required to house the tongue-and-grooved matchboarding); one of the ledgers is positioned near the bottom and one in the mid-area. Two diagonal braces, also reduced in thickness, are positioned on each side of the middle ledger and, ideally, they should be birdsmouth-jointed to the ledgers and the underside of the top rail. Finally, the mid-area is clad with V-jointed matchboarding, which used to be clench-nailed to the ledgers with black-iron cut-clasp nails, but nowadays (for either appearance's sake or speed) are *mistakenly* screwed *(from a longevity point-of-view)*.

Framing anchors: *Figure 48*:   Galvanized-steel framing anchors of various designs for different site-fixings may be used to replace traditionally-nailed fixings in certain places. One example of their use is to attach ceiling joists to timber wall-plates – which were traditionally skew-nailed. The recommended fixings to be used (in every hole of the framing anchor) are 30mm x 3mm (1¼ in. x ⅛ in.) sherardized clout nails.

Freehold property: A lawfully assigned plot of land with a building or dwelling built upon it, that can be held for the duration of the assignees' lives, or inherited by their heirs, or sold to another party.

French doors/French windows: Nowadays, this term usually refers to a pair of fully-glazed double doors – but originally, it referred to a pair of traditional sliding-sash

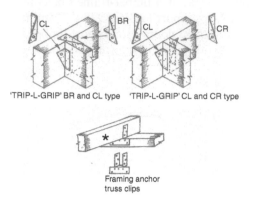

'TRIP-L-GRIP' BR and CL type   'TRIP-L-GRIP' CL and CR type

Framing anchor
truss clips

Figure 48 Floor- and ceiling-joist (*) connections with patented metal Framing anchors

windows at ground-floor or balcony-level, in rooms with very generous storey heights of about 3.353m/11ft. With only a low-rise apron wall beneath the windowsill (about the height of the traditional built-up wainscot/skirting boards), the bottom sash – when lifted up fully – allowed a stooped exit and re-entry over the low-rise apron wall's window sill. In old French chateaus, a French door – or doors – could be found without an apron wall. Instead, there would be a weathered sill to step over.

French drain: Such drains are sometimes used to reduce a high level of water in the ground surrounding a dwelling's subterranean rooms or basement. The drain consists of a certain depth of narrow trench which is dug around the building, against the outer walls. Optionally, before installing the drain, the newly exposed areas of damp wall(s) can be covered with a damp-proof membrane, up to – and no higher than – the dwelling's DPC. Then, 100mm/4 in. diameter, perforated plastic drainpipes are joined up and laid in the centre of the trench (with their perforations uppermost), on a shallow bed of shingle. Ideally, they should be laid with a gradient (of about 1 in 600) towards a small soakaway. Then, the remainder of the trench is filled with not less than 18mm/¾ in. diameter gravel/ballast.

French windows: See **French doors/French windows**

Fresh-air inlet (FAI):   See *Figure 52* on page 142: Such pipes, with a side-vented top (looking like a periscope protruding from the ground) are commonly seen in older properties, near the outgoing inspection chamber, usually up against a low-rise boundary wall. They provide fresh air to the chamber, but are not used nowadays, because venting is done at the head of a drainage system, via a soil-and-vent pipe (SVP) of 100mm/4 in. diameter, that usually rises up against the property's wall, to vent above the head of the highest opening window(s) – by at least 900mm/35½ in.

Frogs: Many manufactured bricks, such as – for example – London Fletton bricks, are produced with an indentation on one of their widest and longest top-surfaces, which is shaped like an inverted, double hip-ended pitched roof. This indentation is called a 'frog' and its purpose is documented as being to stop bricks from sliding about too much on the freshly-laid mortar-bed. However, different opinions exist historically regarding whether bricks should be laid with their 'frogs up', or their 'frogs down'. My own technical/industrial experience (backed up by technical research) convinces me that such indented bricks were originally intended to be laid with their frogs up. Of course, this involves filling in (by trowel) all of the frogs before laying the bedding mortar for the next course of brickwork. So, over time, to avoid this extra mortar (extra work and extra time), the practice of laying bricks with their 'frogs down' developed. Such a technique is definitely speedier. But, on the minus side, because the frogs are only slightly filled with mortar (if laid with their frogs down), such walls are of a less solid construction.

Frost damage: See **Burst water-pipes**

Full measurements: See **Bare, full or tight measurements**

Fungus: See **Dry rot**

Furniture beetle(s):   See **Woodworm**

Furring-pieces/-blocks: See **Firring-pieces/-blocks**

Fuse box: See **Consumer unit**

G

Gabions: These are large, cube-like, welded-wire cages (measuring at least 600 x 600 x 600mm), filled with material such as small or large lumps of natural stone. The gabions make necessarily-flexible *building-blocks* that are used to provide slope-stability and protection against ground-erosion. They are frequently used nowadays instead of traditional retaining walls, in sloping-ground situations close to imminent building structures. And they are reckoned to have a minimum sixty-year lifespan.

Gable: See **Gable-ended walls**

Gable-ended mansard roof: *Figure 49*: This style of roof, in its traditional form, was comprised of two types of roof trusses, built within the confines of gable-ended walls. At wall-plate level (at right-angles to the gable ends) the two steep-pitched sides of the roof were related to *Queen-post trusses*, which allowed habitable rooms to be formed in the roof-void. On top of these, a shallower pitched-roof was formed with *king-post*

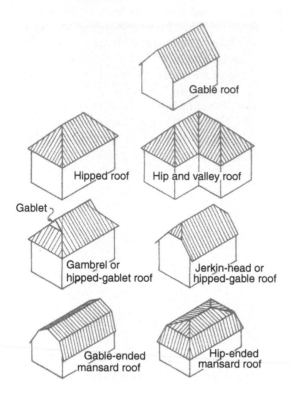

Figure 49 Basic roof designs & rafter-formations

trusses. Modern forms of this attractive, accommodating roof, omit the use of trusses and form the dual-pitched roof (and the interior accommodation) by using steel beams that take their bearings on the gable-ended walls.

Gable-ended walls: These are the walls at the ends of a building, that are built-up higher than the roof's eaves' level, to infill the triangular-shaped area created by the pitched roof.

Gable ladders: *Figure 50*:   Traditionally, when verge projections were required on gable-ended walls, this was achieved by letting the ridge board, the purlins and wall-plates on each side run on and project through the wall by a detailed amount (usually about 150mm/6 in. to 200mm/8 in.), to provide fixings for an opposite pair of common rafters – upon which to fix a pair of barge boards.

However, when truss-rafter assemblies superseded traditional roofing, projecting verges were still required, but the lack of ridge boards and purlins in truss-rafter roofs, presented a fixing-problem. However, it was solved by the idea and provision of framed-assemblies known as *gable ladders*. These pairs of ladders, that require plumb-cuts at their head to support each other – and at their base, for their weak, outer-corners to be supported by the ends of the fascia boards, are fixed on site via nailing to the sides of

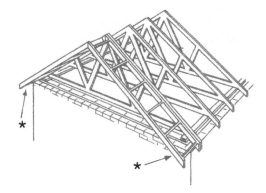

Figure 50 *A pair of gable ladders fixed to a W-shaped roof trass

the first- or last-truss assembly. They are eventually lined on their underside with soffit boards and, on their faces, with barge boards, after being built-in by bricklayers. The usual practice is for the gable walls to be built-up to the approx. underside of the end trusses, then completed (awkwardly) after the ladders have been fixed.

Gable roof: See *Figure 49* on page 122: This design is now widely used in modern roofing because of its simplicity and therefore its relatively lower cost. Another important point worth mentioning, is that the simpler a design, the less roof-junctures there are to cause weathering problems. In this respect, a *Penthouse roof* or a *Pent roof*, with only one surface that slopes in one direction only, might be trouble-free. Such roofs – similar to shallow *lean-to* roofs – are usually built to pitches of 15^0 to 20^0.

Gable-wall restraint: *Figure 51*:   Whether the gable walls are to be partly-built before or after a roof is pitched – and regardless of whether the new roof is of a traditional-type or of modern truss-rafter assemblies, it must be remembered that horizontal restraint straps, of 5mm-thickness galvanized steel, must be fixed at maximum 2m/6ft 6¾ in. centres across the ceiling joists *and* the underside of the rafters. They must be built into the gable walls to help stabilize them, via their L-shaped ends. Note that at the points where restraint straps are to be fixed, across the tops of the ceiling joists and the undersides of the rafters, timber noggings of a substantive sectional-size must be fitted and fixed. And the usual gaps, between the gable-wall's inner surface and the ceiling-joist- and rafter-faces, must be filled with fixed noggings.

Gable window: A window – usually quite small and with a *fixed light* – located in the isosceles-shaped portion of a gable wall.

Galleting: On gable-ended, tiled roofs, this term refers to one or two pieces of broken or cut tile being neatly positioned in the infilled mortar-bedding at the open ends of the

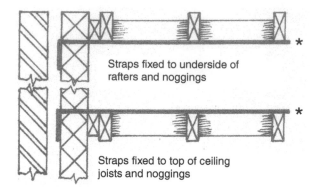

Straps fixed to underside of
rafters and noggings

Straps fixed to top of ceiling
joists and noggings

Figure 51 Gable-wall restraint straps*

first and the last half-round or segmental-shaped ridge-tiles. As the vertically-trowelled mortar in these apex-ends is quite a large lump, prone to sagging, the horizontally-bedded tile-galleting stops the mortar from slumping. Traditionally, small-diameter bottles were used for galleting, with their bullseye-bases flush to the mortar's face – anecdotally with a roofer's note sealed inside.

Gallows brackets: In particular, these are right-angled, triangular-shaped wooden brackets, usually with moulded edges and curved braces, traditionally used to support porch roofs and oriel windows, etc. – as opposed to their ancient use for supporting hung bodies. This triangular shape is also used for various styles of wooden and metal shelf brackets – the metal type sometimes (as with the so-called *London* shelf-*brackets*) minus the diagonal brace.

Gambrel or **hipped-gablet roof:**   See *Figure 49* on page 122: This hip-ended roof displays small design innovations in contrast with basic hipped and gable roofs. In its construction, the ridge board is allowed to protrude at each end to accommodate short (so-called) *cripple* rafters, that have plumb cuts at their heads and compound hip-cuts at their *crippled* feet. Note that these italicized terms are of ancient origin in roofing terminology – and are therefore, hopefully, free from destruction by pc (political correctness).

Gantry: In building terms, this refers to a temporary, storey-height structure (usually of scaffolding nowadays, as opposed to yesteryear's heavy framework of braced dead-shores and timber beams), upon which demountable site-offices were erected. Apart from providing an office and a useful overview of a building site during the critical stages of construction, it sometimes provides useful material-storage space below, when a site is limited in size.

Garden-taps/bibcocks: These taps, often incorporating a portion of threaded nozzle in readiness for the attachment of a hosepipe fitting, are often left exposed on the exterior wall-surface of a dwelling. Yet there are many incidences of such short-lengths of copper supply pipe, that feed garden taps through a wall from a kitchen- or utility room-sink, freezing up and bursting (or causing a compression joint to become detached). Of course, this is only realized once the thaw takes place and can go unnoticed for whatever period of time, causing substantial damage to fitted units and floors, etc.

Therefore, it would be prudent to (a) have an isolating valve fitted internally (near the outlet pipe) for seasonal use, or (b) have a wooden casing (with a sloping, weathered top) built around the pipe and tap – with the encased tap and pipe lagged/packed with insulation.

Garden-wall bond brickwork: Garden walls of dwelling houses are rarely built now-adays, but are commonly seen – usually in a dilapidated state – in the gardens of 18th-/19th-century properties. They are usually built of a one-brick-thickness (225mm/ 9 in.) *Flemish garden-wall bond*, which is often seen to vary in design by the number of stretcher bricks laid in relation to the opposing headers. But the ratio that appears to have been most popular is/was comprised of *one header-brick* to *three double-width stretcher-bricks* laid in each built-up course throughout. However, half-brick thickness garden walls, laid to stretcher bond, with one-brick thickness piers, are seen occasion-ally. But usually built at a later period – and commonly built as low-rise walls to the perimeter boundaries of the public footpath.

Gargoyle: A medieval stone or cast-iron caricature of a grotesque open-mouthed human- or animal-head covering the rainwater shute of a box-gutter outlet through the parapet wall of a box-guttered roof. In practical terms, this primitive building-component threw the rainwater clear of the building, without the need for a hopper head and rainwater downpipe. But allegedly, the grotesque caricatures were intended to keep out evil spirits.

Garnet hinges: See **Tee hinges**

Gathering-over: See **Corbelling**

Gauge: In building-terminology, this reference to measurement refers to: **1)** The ratio of a mortar-mix, i.e., 3:1 (three of sand to one of cement); **2)** The repetitive measurement (spacing) of tiling battens, etc.; **3)** A joiner's tool for gauging/marking predetermined cutting- or planing-widths on timber.

Gauged arches: In good-quality brickwork, geometrically-shaped arches that are seg-mental, semi-elliptical, semi-circular, etc., are usually built with taper-shaped, *gauged* facing-bricks called voussoirs – as opposed to being built with ordinary, non-taper-shaped facing-bricks that are used on cheaper work (or *common* work to be plastered). Note that such arches as these have *tapered mortar-joints*, which are technically (and aesthetically) frowned upon.

Gauged bricks: See **Voussoirs**

Gauge rod: See **Storey rod and Gauge rod**

Gauging box: Traditionally, this was a bottomless wooden box, with side-projecting handles, which was used to measure separate, equal amounts of sand or cement to a given ratio. On a larger scale, mortar and/or concrete are now gauged and mixed by machinery – and on a smaller scale, buckets, etc. can be used for gauging.

Gazebo: A small open-fronted outbuilding, with a park-like, fixed garden-seat, usually facing an attractive view, which – given a fine day – one might choose to *gaze* (*gaze*bo) out upon.

General Foreman: See **Site Agent**

Geometrical staircase: See *Figures 25 and 26* on pages 63 and 64: A curved staircase designed to fit a circular- or elliptical-shaped wall, usually against the concaved surface, but it can also be against the convex surface. The tapered treads of a geometrical stair either radiate from a determined centre on the concaved side of the curved wall or from a similar centre on the convex side – depending on which side of the wall the stair is to be fitted. On either side, the position of the radiating centre (required to set out the tapered treads to meet AD K1's Building Regulations' requirements) is a *geometric normal* tangential to the wall's curve. So, when setting out the tapered steps, trial-and-error radiating movements will depict their allowed-positions to meet the restraints of the TR+G formula in relation to *minimum going* and *maximum rise*. Geometrical stairs require so-called *continuous wreathed-strings* and *wreathed-handrailing*, usually without newel posts. Such a potentially weak balustrade gains its strength (rigidity) from a more detailed (stronger) fixing of the balusters at their extremities. Note that the *outer string* of such staircases can have parallel edges (like the outer-string boards of common stairs), or they can be of a cut-and-bracketed stair design.

Georgian baluster-sticks: This description is a reference to unusual-shaped baluster *sticks* (i.e., of a square, sectional shape), which have a small portion at the top and the bottom that remains square (similar to lathe-turned balusters) – but (unlike the latter), the ornate pattern in the central portion is moulded *flatly* across the grain on each of the four sides. Thereby, still retaining the baluster's squareness throughout its entire moulded and unmoulded length. Such attractive balusters are usually only seen in early-18th- to early-19th-century properties. Usually, the newel posts and the ornate caps to such stairs are similarly moulded.

Georgian wired-plate (GWP) glass: Of 6mm/¼ in. thickness, with a square-patterned wire-mesh embedded within it – to resist disintegration in a fire. GWP glass can be obscured or plain and it has a fire-resistance (FR) of 30 minutes.

Girder: See *Figure 36*: This is a generic term which is usually applied to a large-sectioned beam that supports floor joists running at right-angles above it. Historically, the girder would have been of timber or steel, but nowadays girders are usually of steel – and they are placed across the shortest span. Such floors can be technically referred to as *double floors*. From a structural-design point-of-view, they are used when the shortest span is too wide to enable the use of standard-sectioned timber floor joists. Note that the projecting underside portions of steel girders need to be encased to protect them from fire. Traditionally, this was done with timber cradling and plasterboard linings.

Glazed bricks: Such bricks, comprised of pressed clay, will have been dipped in prepared glazing, after their face-sides have been brushed with white clay and dipped again in the same coating. They are then burnt in kilns at a high temperature. They are/were used in places that required to be kept clean by frequent washing, such as laboratories, restaurant kitchens, larders, public toilets, etc.

Glazing bars: See *Figure 17* on page 42: When a window or glazed-door aperture is designed to be divided up into smaller panes of glass, the rebated, T-shaped dividing-members are technically referred to as glazing bars. And, because the practicable size of the rebates is limited to receive only single panes of glass, when double-glazing is required to be installed, the timber-thickness of the window or door must be increased – and this increase beyond the traditional norm is also required to meet a panelled, insulated door's upgraded energy rating in compliance with the Building Regulations' AD, Part L1B. So, with regard to double-glazed sealed-units, if glazing bars are required, they have to be imitation, usually located within the sealed units.

Glazing sprigs: These must be used sparingly to provide necessary support to putty-embedded panes of glass, after the glass has been firmly pushed/embedded against the back-putties. To avoid cracks, sprigs must not be placed too near the end-corners of the glass – and they must be driven in low enough to be covered by the bevelled face-putties. Panel pins are sometimes used instead of glazing sprigs, but their protruding tops often scratch the glass as they are being driven in, thereby creating another potential crack. Sprigs are about 12mm/½ in. long and have a minute square-top, tapering down to a point.

Glue: See **Woodworking adhesives**

Glue-lam worktops, etc., and engineered timber: Gluelam (glued and laminated) construction (also referred to nowadays as *engineered wood*) uses any number of narrow strips of timber (laminae) – of softwood or hardwood – with their side edges glued and laminated together to build up whatever width or length of board is required. Providing the appearance of the multiple edge-joints and their staggered end-joints is acceptable on a particular job, this is an excellent way of building up and stabilizing wide surfaces and any length of board. Technically, the end-grain growth-rings on the laminae, that can be seen to radiate one way or the other in relation to the timber's face-side surface, should be arranged to oppose each other – to inhibit surface-cupping – but one rarely sees this time-consuming consideration done commercially.

Glue line: The line of glue in the abutting surfaces of timber joints; and such joints are reckoned to be stronger if the glue line is longer – as in a multi-fingered bridle joint.

Going: See *Figure 73* on page 253: The horizontal dimension from the nosing edge of a stair-tread to the nosing edge of the next consecutive tread above it.

Gothic arch: 1) Geometrically, such an arch (that resembles a spear-head) can be an *equilateral Gothic arch*, whereby the radius needed to form the arch-shape is equal to the arch's span – and is radiused from each side of the arch's base (the springing line); or, **2)** it can be a so-called *depressed Gothic arch* (also referred to as an *obtuse-* or *drop-Gothic arch*) whereby the two radius points (centres) on the springing line, are dictated by a lower rise chosen on the centre line; or, **3)** it can be a *lancet Gothic arch*, with a higher-pointed spear-head, created by the two radiating centre-points being set up to a chosen amount on each side of the springing line, *outside* of the required (or chosen) arch-span.

Grade I and Grade II listed buildings: A Grade I listing will usually have been awarded to buildings thought to be of *exceptional interest* – and sometimes thought to be of international importance. To be listed as Grade II (a more ommon listing than Grade I), a building must be of national importance and of *special interest*. Over 95% of listed buildings are Grade II, compared to the special rarity of a Grade I listing.

Listed buildings are subject to strict regulations to protect their architectural and historic significance. Alterations and/or building work should not be carried out without prior written consent from the local authorities.

On Grade II listed buildings, repairs-only can be carried out without prior consent, but should be made with like-for-like materials and traditional methods of application. A known issue of annoyance for listed-building owners and conservationists is that double glazing is not allowed.

Graining: Painting and *graining* material-surfaces to look like wood or marble, etc.

Gravel boards: See **Featheredge-boarded fence**

Green brickwork, etc.:   This is a site reference to freshly- or recently-laid brickwork (the mortar thereof) or concrete, etc., inferring that these cementitious components have not fully *set* (hardened by their chemical process) and are still in a non-approachable, delicate state.

Grinning: This trade reference refers to inferior or skimped paintwork, where amateurish or insufficient coats of paint have been applied, causing a blotchy appearance to *grin through* to the finished surface.

Ground floor: See **Numbering and naming of floor levels in the UK and the USA**

Grounds: See **Timber grounds**

GRP (glass-reinforced plastic) flat roofs: This roofing material is also referred to as being of fibreglass, because it is comprised of a resin with glass or fibrous strands that reinforce it. Such roofing is not usually regarded as a job for a DIYer, because the process involves first applying a layer of resin, followed by laying a mat of fibreglass strands – then another layer of resin, before a final top coat is applied.

GRP (glass-reinforced plastic) valley liners: Seemingly, the most popular way to form roof-valleys nowadays is to use a preformed GRP valley liner. They are available with: **a)** an inverted Vee-shaped upstand moulded along their central length – as an abutment-guide for splay-cut slates to be laid tight up against them, on each side of a dry-valley system, or **b)** with a factory-bonded sanded-strip running down each side of the valley liner – to allow good adhesion of the mortar-bedded and side-edge pointed splay-cut tiles on each side of the valley.

Grub screw: A very small, headless, screw-slot topped, screw-threaded connector, used to fix a traditional door knob to the threaded hole in the metal spindle.

Gunstock joint: See **Diminishing stile/gunstock joint**

Gusset plate: A triangular-shaped metal (or graded plywood) face-plate joint-connector.

Gutters: See **Rainwater gutters**

GWP glass: See **Georgian wired-plate (GWP) glass**

Gyproc plasterboard (from **British Gypsum/***Saint-Gobain***):** The so-called *building-board/Wallboard*, made for lining ceilings and walls, is comprised of a core material of anhydrous plaster or gypsum, sandwiched between thick, heavy-duty paper that also covers the long edges. The boards' superficial-sizes and thicknesses are: **1)** 900mm x 1800mm and 1.2m x 2.4m x *9.5mm thickness*, both available with tapered- or square-edges (the former for dry-lining and flush, face-fillings, reinforced with self-adhesive reinforcing scrim; the latter for 3mm open-jointed abutments and plaster-skimmed finished surfaces; **2)** 900mm x 1800mm and 900mm x 2.4m *x 12.5mm thickness*, both available with tapered- or square-edges. **3)** 1.2m x 2.4m., or 1.2m x 2.5m., or 1.2m x 2.7m x *12.5mm thickness* and available with tapered- or square-edges. **4)** 1.2m x 3.6m x *12.5mm thickness*, but with tapered edges only. **5)** 900mm x 1800mm., or 1.2m x 3.0m x *15mm thickness*, with tapered edges only; **6)** 1.2m x 2.4m and 1.2m x 2.7m *x 15mm thickness*, with tapered- or square-edges.

Note that: **1) Gyproc WallBoard** (previously referred to as *plasterboard*), is British Gypsum's standard gypsum plasterboard, which is recognizable by a *buff*-coloured paper encasement; **2) Gyproc Moisture Resistant** plasterboard, is gypsum plasterboard with moisture-resistant additives in its core, which is recognizable by a *green*-coloured paper encasement; **3) Gyproc FireLine1**, is gypsum plasterboard with fire-resistant additives, which is recognizable by a *red*-coloured paper encasement; **4) Gyproc FireLine2**, has the same characteristics as **FireLine1**, but with so-called ACTIV*air* technology; **5) Gyproc SoundBloc** is gypsum plasterboard with a high-density core, for enhanced sound-insulation, which is recognizable by a blue-coloured paper encasement; Note that Gyproc SoundBloc1 boards are also available in a Moisture Resistant (MR) version, for specifying in intermittent wet-use areas. They are also available with ACTIV *air* technology; **6) Gyproc SoundBlocF** is of a gypsum core, with fire-resistant additives and a high-density core for enhanced sound-insulation performance; it is also recognizable by a blue-coloured paper encasement; **7) Gyproc Plank** is a standard gypsum plasterboard, located as an inner layer, which has a green coloured-paper encasement.

Gypsum laths: These originated in 1910 as an alternative to wooden-laths-and-plaster and were originally referred to as 'Rock laths'; they were widely used from the early 1930s and preceded modern-day Gyproc wallboard/plasterboard. They were 450mm/ 18 in. wide x 1.2m/4 ft. long x 10mm/⅜ in. thick. The long, paper-bound edges on each side were segmental-shaped, so that when they were fixed side-by-side, with a minimum 4mm/⅛ in. gap between them, the initial skimmed caulking plaster squeezed easily through the gaps to form strong, mushroom-shaped 'locking keys' prior to the application of the overall setting/skimming coat. No reinforcing scrim was used – seemingly without adverse consequences. Note that the paper-bound, segmental-shaped edges ran at right-angles to the ceiling joists – and the square, unbound edges of the boards were fixed centrally on the joists and staggered in relation to each row of boarding

H

Half-brick wall: This is a reference to a *single-leaf* (or *skin*), *stretcher-bond* wall with a theoretical thickness of 114mm/4½ in. – and is therefore similar to the *outer-skin* of a *double-leaf* cavity wall. Such single-skin walls, if built independently, to build a shed or a garage, for example, will need to be stabilized by integrated, intermediate brick-piers and quoins. Note that half-brick-thickness walls do not meet the Building Regulations' criteria as being thick enough – on their own – as exterior walls on any habitable part of a dwelling.

Half-landing: A landing platform between two flights of stairs, that breaks the journey and turns the top flight (and the user) around by a *half-circle* (180⁰), hence the *half-landing* term.

Half-lap joint: A traditional carpentry joint in roofing, used for jointing and joining timber wall-plates lengthwise, or at right-angles on hipped ends. The length of lap should equal the width of the sawn-timber plate, which is normally 100mm/4 in.

Half-round ended step: This is sometimes referred to as a *D-ended step* and is a reference to the bottom step of a stair-flight that protrudes beyond the front- and side-faces of the newel post. The protruding step is semi-circular shaped and is housed into the outer, side-face of the post. As well as being a decorative feature, the step's main, traditional purpose was to increase the stability of the newel post, by virtue of the string's tenons being raised up from newel-entry at the side of the bottom riser to newel-entry at the side of the second riser.

Handed doors or **windows:** Whether doors and/or windows open on the left or the right-hand side is a design feature, but occasionally, certain door- or window-furniture has to be ordered as left- or right-hand fitting, for varying reasons. For example, *rising-butt hinges* – sometimes required to lift a door over an irregular- or a carpeted-floor, are made to spiral differently according to whether a door is right-hand- or left-hand-opening. Different hands of rising-butts are marked R/H and L/H. So, a simple rule for determining this, is to imagine facing a door as it will open *away* from you, and if you want it to open on its left side, then you require left-hand rising-butt hinges – and vice versa tells you to order right-hand butts. This definition-rule, if required, also works for windows. However, confusingly, in the UK, these trade-definitions contradict *descriptive references-of-opening*; i.e., a right-handed opening door (hung on the right-hand side) is one that is opened towards you – which (remember the above rule) would (if required) be hanging on L/H rising butts!

Handrail bolts: Traditional handrail bolts, of 10mm/⅜ in. diameter, were threaded to take a square nut at one end and a circular, side-notched nut at the other. When joining the ends of two handrails (or the end of a handrail to the end of a handrail-wreath), central holes were drilled into the ends of each; square, stubbed-mortices were made on the underside to accommodate the nuts – and, to inhibit the joint from rotating when it was being glued and the nuts were being tightened, two small, blind holes were drilled into the end grain on each side of the bolt holes, to receive 6mm/¼ in. wooden dowels.

Handrail knee: See Mitred knee

Handrails: As per the updated Building Regulations, AD (Approved Document) K1, stairs should have a handrail on at least one side if they are less than 1m wide and, if the stair is wider, there should be a handrail on both sides. Handrails should also be provided beside the two bottom steps in public buildings and where stairs are intended to be used by people with disabilities. In all buildings, handrail heights should be between 900mm and 1m., measured vertically above the pitch line to the top of the handrail – or from the landing to the top of the handrail. Usually, the minimum of these extremes applies to the pitch line – and the maximum to the landings. For buildings other than dwellings, handrail heights should be between 900mm to 1m above the pitch line to the top of the handrails – and between 900mm to 1.1m above finished floor levels and landings. Note that, only for buildings other than dwellings, AD K1, diagram 1.13, illustrates a design-shape and size of handrails to be used. Also, for this category of building, regulation 15.5.2 states that, if a secondary, lower handrail is required (for children), it should be at 600mm above the pitch line.

Although the stair regulations are not completely covered here, I will end with their reference to *Buildings other than dwellings*, where AD K1/paragraph 1.36(f) and (g) infers that the ends of handrails (if not built into newel posts) should be finished in a way that reduces the risk of people catching their *clothing* on the end projection(s). This is a welcome *partial* return of a previous lapsed stair-regulation in the 1976 Building Regulations, which stated that such open-ended handrails should be terminated by a scroll or other suitable means, to 'reduce the risk of people catching their *bodies*' (and, less importantly, *clothing*, perhaps) on the end projections. Of course, wooden scrolls are too elaborate nowadays, but a quadrant-shaped handrail wreath could be/has been used to return a projecting handrail into/up against the face of a wall.

Handrail scroll: At the base of a geometrical stair, there is usually a *curtail step*, which protrudes beyond the outer stair-string in the shape of a spiral-coiled scroll – and this geometrically-developed shape exactly replicates the shape of the geometrically-developed handrail scroll of the handrailing immediately above. Traditionally, the main lateral support for the scroll (and its bolt-attached handrail) was gained by a centrally-positioned, ornate, cast-iron baluster that rose up from a centrally-drilled mortise hole in the geometric-centre of the protruding curtail step. This was enhanced by closely-spaced wooden balusters that were fitted and fixed around the step, between the upper and lower shapes of the two scrolls.

Handrail wreaths: See *Figure 25* on page 63: This refers to certain parts of continuous, geometrical (curved and twisted) handrailing, where it must relate to radiused turns or round-ended bottom steps. Handrail wreaths, therefore, might best be simply described as twisted pieces of handrail that conform to *helical twists or coils following the ascent or descent of the stair-strings below*.

Traditionally, these wreaths were made by hand (and, still are to some extent), with the essential aid of a **face mould**. This is a thin plywood template of 6mm/¼ in. thickness, that is developed geometrically from information gleaned from the stair-designer's drawing. The salient details required are whether a single-twist or a double-twist wreath is indicated, the size and shape of the handrailing and the pitch angle of the stair. From this, the wreath-maker makes a full-size localized drawing of these details, from which a complex piece of handrailing-geometry is carried out to determine the wreath's basic shape and thickness required. This shape is also the boomerang-like shape of the face mould needed. So, this is carefully cut out and pasted onto the plywood, to enable the shaped face mould to be made. When ready, it is laid tangentially-to-the-grain on a very thick piece of selected timber, and marked around its edges for initial cutting-to-shape on a bandsaw. The procedure after this to produce the twisted wreath is not too complex, but is a skilled job and requires at least sketch drawings to explain it.

Hang/hanging: To fit/hang doors, or wooden casement-windows, etc., by their hinges.

Hangers: See **Joist-hangers**

Hanging/hung tiles: These commonly-used terms refer to plain (non-profiled) roofing tiles of various materials, such as clay, concrete, slate, cedarwood, Eternite, etc., that are *hung* and nailed to horizontal tiling/slating battens that have been fixed to the vertical substrate(s) of a building. Historically, hung tiles were also sometimes hung on the battens and bedded onto the substrate with mortar, as opposed to being nailed to the tiling/slating battens. This practice no doubt eliminated the *chattering* noises alleged to be heard on a windy night and no doubt reduced the number of tile-breakages caused by wind-suction.

Hardwood: This term is a commercial description for the timbers used in industry, which have usually been converted from broadleaved deciduous trees, belonging to a botanical group known as angiosperms. Occasionally, the term hardwood is contradictory to the actual density and weight of a particular species. For example, balsawood is a hardwood which is of a lighter weight and density than most so-called softwoods. The most desirable hardwoods for exterior use in the UK are oak and teak – two very much depleted species.

Hasp and staple: A traditional locking-devise used on shed doors, boarded-gates, tool boxes, etc., that requires a separate padlock. More substantial, higher quality hasp-and-staple fittings are available for industrial-type framed, ledged and braced (FL&B) doors, requiring greater security.

Haunch: In joinery terms, this refers to the reduced portion of a mortised-tenon that projects from a tenon's shoulder at the end of a right-angled junction, to (a) create a part-mortised tenon instead of a bridle (comb) joint and (b) to establish a part-haunched housing in order to eliminate the risk of *cupping* via an inferior butt-joint.

Haunching: In traditional glazed-earthenware drain-laying, this is a reference to the banked-up concrete laid at the sides of newly-laid pipes to support them above the bedding concrete.

Head: This is the identifying name given to the top, horizontal parts of a window-frame, or a doorframe. So, a doorframe might consist of two rebated and ovolo-moulded jambs – and one rebated and ovolo-moulded *head*. Contradictorily, in joinery terms, the top, horizontal parts of a casement-sash, or a panelled door, are referred to as *top rails*.

Header brick: In brick-bonding terminology, this is a reference to the narrower, end-face of a brick.

Headroom (above floors, stairs and landings):   Headroom between floor-surfaces and the underside of ceilings is covered herein under the heading of **Storey-heights**, which is tied up with a Building Regulations' definition, related to structural considerations, however, such consideration only seems to be relevant if ground floor story-heights are above 2.7m/8ft.10⅜ in. Above ground-floor level, the storey height for other floors is reckoned to be the same, but the 2.7m is applied differently – it is said to relate from the ceiling at ground-floor level to the underside of the ceiling at first-floor level. Thus, making the headroom of subsequent storeys 2.7m minus a floor-thickness of, say, 270mm = 2.430m/7ft. 11⅝ in.

   The minimum regulatory headroom above the theoretical pitch-line (resting on the nosings of all the steps) should be not less than 2m measured vertically at any point to the underside (soffit) of the stair above, or to the bottom corner-edge of the bulkhead of the landing above. For loft conversions, where there is not enough space to establish this height, the headroom will be satisfactory if the height at the centre of the stair-width is 1.9m, reducing (if need be) to 1.8m at the side of the stair, against the wall.

   Headroom above landings is also not less than 2m above the theoretical pitch line.

Hearth/hearthstone: The immediate floor-area in front of a fireplace, which must be of non-combustible material such as stone, quarry tiles, sand-and-cement screed, etc. – and (to meet Part J of the Building Regulations) hearths should project not less than 500mm/19¾" from the chimney-breast front and not less than 150mm/6" from the side-edges of the fireplace-opening.

Heart-shakes: With reference to timber conversion, heart-shakes are splits, caused by shrinkage, that radiate from the pith (central area) of an over-mature tree. They are only discovered after a tree is felled and reduced to a bole (tree trunk).

Heartwood: The mature, central portion of a tree, which, because of the effect of its stabilized substances such as gum, resin and tannin, is usually darker than the sap-wood – and usually more reliable for retaining its machined shape and size.

Heat pump: This is a relatively modern heating system that uses an air-sourced heat pump to heat circulated water for a property's heating system, as well as provide domestic hot water. It consists of an evaporator unit fixed on the exterior of a building, which produces a liquid- vapour mixture. This passes into the building, via a compressor, that converts it into a high-pressured vaporized refrigerant within a condenser unit. Then the cooler air in the building condenses the vapour which converts it into a heat-source.

Heave: See **Tree-root damage to dwellings**

Heel-thumping test: This is a non-intrusive test used on suspended timber floors, to gain an impression of their soundness. On aged property, this test is of such floors commonly found at ground-floor level, which quite often sound and 'feel' springy or spongy, which indicates the likelihood of wet- or dry-rot decay – especially to the ends of the joist-bearing wall-plates. Note that when exposed floorboards show signs of decay (from wet-rot, dry-rot), or show small holes from an attack by the common *furniture beetle*, (also referred to as *woodworm*), it may not be wise to do too much heel-thumping.

Helical hinges: These traditional hinges, also called *bomber-butts* (because of their double-cylindrical appearance being reckoned to resemble a pair of high-explosive bombs), were used on double swing-doors – this either being one door or a pair of doors – as a cheaper alternative to more expensive floor-spring-and-top-pivot hinges. The latter required to be housed in the floor, with a pivot at the top, housed into the underside of the frame-head.

Herringbone strutting: This traditional method of *strutting* (reinforcing the stability of on-edge floor-joists), using 38 x 38mm/1½ x 1½ in. or 50 x 50mm/2 x 2 in. sawn timber, although still effective and occasionally used, nowadays has to compete for speed with struts made of lightweight steel. The method of fixing starts by marking a line across the joists (usually in the centre of the span, as detailed under **Solid strutting**); from this, pencil lines are squared down the sides of the joists and – in the case of timber strutting – another line is struck on top (with a chalk line), parallel to the first and set apart by the joist-depth minus 20 to 25mm/¾ to 1 in. Then, the strutting timber is laid diagonally within these two lines and marked underneath (from above) to produce the required plumb cuts (vertical faces of the angle). Cutting and fixing the struts are done in a kneeling position from above, using 50 to 63mm/2 to 2½ in. round-head wire nails.

Prior to fixing the struts, the joists running along the sides of each opposite wall will need to be packed – with wedges, traditionally, but timber-shims can be used – and nailed (or adhered), directly behind the line of intended struts. If interested, see: **Steel herringbone-struts**.

High-alumina cement (HAC) and its problems: Historically, *high-alumina cement* was sometimes used to speed up the chemical setting-action of the concrete in the manufacture of precast-concrete units, thus achieving a better production rate. The mould-box sides and ends (of cast-beams or floor-panels, etc.) could be carefully removed earlier-than-normal from the factory-made units, cleaned up, oiled and quickly reused. But, in early 1974, two precast concrete roof-beams collapsed in a school in Stepney, London; which led to an urgent investigation by BRE (the Building Research Establishment). When it was discovered that high-alumina cement was used in the production of the beams, the use of HAC was stopped and became the subject of a BRE Current Paper: CP 34/75 High-alumina Cement Concrete in Buildings, published in 1975.

According to this, the use of high-alumina cement can cause a chemical change in the concrete, with the likelihood of structural failure to pre-stressed and pre-cast concrete floor-beams, if their span is in excess of 5m/16ft 5 in. If the spans are less (or not more than 5m.), the risk of failure is thought to be very small. However, it is believed that high alumina cement has not been used in concrete since the Stepney school incident. From a surveying point-of-view, one should be vigilant for signs of stress in any viewable precast-concrete units seen in buildings – regardless of their age or whether HAC was a constituent part.

High-level WC cistern: This traditional, high-level, siphonic flushing cistern (usually fixed at about 2m/6ft. 6 in. above the floor), with its side-hanging pull-chain, is still very much in use – usually in a Victorian-period form, with a classic-looking cistern seated on ornate metal brackets and a 32mm/1¼ in. diameter, chrome flush pipe. When the chain is pulled, the lever lifts the diaphragm plunger and triggers off the siphonic action that releases the water. It is believed that this cistern's maintenance-free and non-water-wasting performance outstrips many modern dual-flush cisterns – as explained herein, under: **Water closet (WC).**

Hinge-bound doors: Doors hung within a doorframe or door-lining (the latter being more common) require a tolerance-gap around them on all four edges. These gaps are called *joints* and their sizes should be 2 to 3mm at the sides and top – and 3 to 6mm at the bottom. If the hinges have been sunk-in too much and the required joint is non-existent, the door will be *hinge-bound* and will not close properly until the hinges are packed out with cardboard, etc.

Hinges: A wide variety of hinges are used in building- and refurbishment-work, but the most common – historically and nowadays – are butt hinges. And the common sizes used (with reference to their length) are either 75mm/3 in. and/or 100mm/4 in. What does seem to have changed, in recent times, is that mild-steel butts (requiring to be painted) appear to have been superseded by maintenance-free finishes, such as stainless-steel (satin or polished) and brass (polished, electroplated or antique), etc. They are also rated for fire doors and can be fitted with intumescent pads.

Hip: This term refers to the *dihedral-angle-edged* timber board that separates the two, equally-inclined roof-surfaces of a hip-ended roof.

Hip-and-valley roof: See *Figure 49* on page 122: This is a reference to a pitched roof on an L-shaped building design, that may have hipped or gable ends, but at the building's right-angled juncture, there will be only one hip rafter immediately opposite a valley rafter.

Hip-ended mansard roof: See *Figure 49* on page 122: This style of traditional roof is similar to the mansard roof described above under **Gable-ended mansard roof**, except that it has hipped ends to the lower steep-slopes and hipped ends to the upper roof's shallow slopes. This chalet-style property is sometimes seen with dormer windows to the lower, steeper slopes.

Hip iron: A scroll-ended metal bar fixed to the top surface of a hip rafter at its bottom end, to help support the lowest, mortar-bedded hip tile. Nowadays, they are usually galvanized.

Hipped ends: See *Figure 49* on page 122: Technically, this refers to a roof with two pitched sides, say, at 42^0, that has two return ends that are also pitched at 42^0. Such a four-sided formation creates two hipped ends.

Hipped-gable or **Jerkin-head roof:**   See *Figure 49* on page 122: This style of traditional pitched roof is comprised of gable-ends that do not rise to an arrow-head point (apex), but which are levelled off at a point that equates to about ¾ of the roof-rise. At this point, short-length wall plates are bedded and small hip-ends are formed. The historic name of *jerkin*-head seems to be linked to the very imaginative vision of a traditional sleeveless jacket.

Hipped roof: See *Figure 49* on page 122: This term describes the most commonly-used arrangement of two hipped-ends on a rectangular-shaped building. And because the four roof-slopes are pitched at the same angle, this geometrically results in the four hip rafters being at 45^0 in orthographic plan-view. It also creates a shorter length of ridge-board area – but a large amount of diminishing jack rafters (supporting the hips), which require compound edge- and plumb-cuts and a variety of diminishing lengths. The roofer's work is also more involved, with multiple cutting of tiles or slates to fit the acute angles of the eight hip-edge abutments. Therefore, although this style of roof might be considered attractive, it is usually much more expensive.

Hip rafters: Usually of relatively thin sawn timbers (25mm/1 in. to 38mm/1½ in. thick) with widths according to the roof-pitch (the steeper the pitch, the deeper the hip-rafter boards), which are set-up at the 45° dihedral angles of a hip-ended roof.

Hoarding: A temporary fence put around a building site, usually with locked gates for vehicular entry and exit and separate locked doorway(s) for site personnel and visitors. Nowadays, the framed plywood or OSB (oriented strand board)-clad panels used for hoardings are usually prefabricated and delivered to the site for erection; traditionally, the hoarding was made and erected wholly onsite.

Hogsback-tiled valley: Such valleys can still be seen in existence, so some knowledge of them might be useful when evaluating descending-damp problems. Visually, they can be detected by the appearance of a rather narrow *trough*-gutter (of about 125mm/5 in. width) sunken into the inverted valley junction, below the edges of the cut roof-tiles bearing on the tiled-valley's side-edges.

The end-collar-jointed hogsback tile-lengths – which have a wide, shallow, inverted parabolic-arch shape, are mortar-bedded onto three built-up valley boards (one furred-up at the base of the valley, the other two being splay-jointed on each of the sloping roof-sides). However, note that the three built-up valley boards should have been treated like a lead-lined gutter – and lined with overlapped sheet-lead before the hogsback valley-tiles were laid.

Whether a traditional roofing-felt membrane was laid over the rafters (and lapped onto the welted lead-edges of the hogsback valley), will depend on the age of a property. Prior to the early 1930s, roofing felt did not exist. Nowadays, breathable, impervious membranes (such as Tyvec Supro or Marley Supro) would need to be draped over the whole skeletal roof structure, beneath the slating- or tiling-battens.

Homebuyer Report: See **Building Inspections**

Honeycombed sleeper-walls: See **Sleeper walls**

Honeycomb roof-slating: See **Asbestos-cement roof-slates**

Hook-and-band hinges: These types of heavy-duty wrought-iron hinges can be used on wooden garage- or workshop-doors, etc. Fixed by coach bolts and screws – with an option of hooks that can be built into masonry, instead of being screwed onto the door/ gate jambs. The bands vary in length between 450mm/18 in. to 760mm/30 in.

Hopper head: A flared-out, open-topped box-shaped fitting at the head of a lead, cast-iron or plastic rainwater pipe, used to receive rainwater from flat-roof chutes, or from small-diameter waste pipes, etc.

Horizontal cracks in brickwork joints: See **Wall-tie corrosion and expansion**

Horn(s): This term refers to timber *heads* and *sills* of traditional windows and/or doorframes, whereby small extensions of about 75mm/3 in. were added beyond their required width to facilitate effective mortise-and-tenoning (M&T) of the joints formed at their ends. The horns, in fact, were used for three reasons: 1) To counteract the sheer-force effect when the wedges were driven in between the M&T joints at their outer ends; 2) To protect the corners and sides of the frames during transport to the site; 3) Bricklayers used to leave the top horns on and build them into the brickwork to hold the window or doorframe in its position. (This was also when bricklayers and the industry used the wooden heads as supporting-lintels!)

Houses in Multiple Occupation (HMOs): See **Flat-conversions**

Housings: Shallow trenches or grooves, usually cut across the grain of timber to receive (house) another piece of timber and so form a right-angled formation – such as a door-lining housed into a door-lining head.

Hot-air heater: This term usually refers to a central-heating system that uses electricity to operate a fan-assisted heater. This can be programmed to blow heated air through a ducted system serving floor- or wall-grilled outlets.

Hot water cylinder: In a traditional hot-water supply system, this is used to hold a ready-supply of hot water. It may be a *direct* or *indirect* cylinder, dependent upon a heat-exchanger being fitted. If one is not fitted, then the hot-water vessel is an indirect type of cylinder.

Hung tiles: Tiles of clay, concrete, slate – or cedarwood shingles – are traditionally hung and fixed to horizontal battens/grounds that are fixed to a substrate of solid blockwork, brickwork or to timber studwork, as an elevational façade. Hung tiles of clay or concrete can also be found to be bedded to the substrate with cement-mortar – such being a practice of yesteryear – but which was done to combat breakages via wind-suction.

Hybrid flat roofs: As this term suggests, this type of roof refers to certain constructions which are mixed and do not fall within the warm- or cold-deck categories. This is because some structural decks are themselves comprised of insulating materials, such as woodwool or, in other cases, insulation is added above the deck, in addition to insulation at ceiling level.

Hydrostatic pressure: In building construction, this scientific term relates to the increased lateral-pressure of water that builds up at or near the base of – for example, **1)** retaining walls with higher ground up against them on one side only, requiring them (when being built) to have escape outlets (weep holes) built into the wall's base, usually in the form of small-diameter, wall-width pipes spaced at varying centres according to the height of the retained ground; or **2)** timber formwork (shuttering) to a reinforced-concrete column, which may require the four-sided column-cramps to be closer together near the column's base to combat the hydrostatic pressure of the initial fluid concrete.

Hygroscope: A mobile-phone-sized instrument that displays, without measuring, changes in the humidity of the air.

Hygroscopic soot-ash damage: See **Deliquescent soot ash staining to chimney-breast plaster**

I

Immersion heater: A twin-pronged (or double twin-pronged) electric tube-heater, permanently submerged in a copper hot-water cylinder and controlled by a thermostat.

Indenting: See **Toothing and block-bonding**

Independent scaffold: This tubular-alloy scaffolding, built as an independent structure, is regarded as an essential provision for medium-to-large new-build contracts – as opposed to the use of a *putlog scaffold*, described herein under that heading. But, like the putlog scaffold, it is still erected in incremental stages of 1.52m/5 ft. *lifts*, as and when the newly-laid brickwork has reached that workable height – and has *set/hardened* enough (usually overnight) to withstand being built-up again. This self-supporting scaffold, that is gradually built-up as each workable height of brickwork has been laid, is built in the form of a series of ladder-like structures. They are each comprised of two upright tubular *standards*, with short, horizontal scaffold-tubes (putlogs) fixed across them, via bolted clips. Below the standards and up against the putlogs on each side of the ladder-like frames, long, horizontal ledgers are clipped into position. Initially, on the first lift of scaffolding above ground level, relatively short-length tubes will be swivel-clipped diagonally as braces across the scaffold's narrow width – and eventually, as the scaffold lifts increase in the wake of the brickwork lifts, more short-length braces across the scaffold's width – and longer, diagonal braces across the scaffold's frontage will need to be added. If considered necessary, raking outrigger scaffold-tubes are sometimes added to hold the scaffolding against the building.

Indirect hot-water supply system: Traditional hot water systems used a so-called *direct cylinder*, whereby the water passed directly from the boiler to the hot-water cylinder, ready to be drawn off at the taps. But, in *hard water* districts, these cylinders *furred up*, so to prevent this happening, *indirect cylinders* evolved. These are fitted with coil-type or annulus-type heat exchangers. For comparison, see **Direct hot-water supply system**, written herein.

Inglenook fireplace: This Scots/Gaelic term (seemingly meaning *angled-corner* fire-place) refers to a wide, recessed, low height alcove, housing a recessed open fireplace in its mid-area. The recessed *nook* (corner) on each side is recorded to have been used for housing a seat, for relaxing by the fireside. Supporting the low-height alcove's brick wall above the low opening, one often sees an ancient-looking oak *bressummer* beam.

Inner skin (leaf):   This is the wall built on the dwelling-side of a cavity wall, tradition-ally constructed of brickwork, but nowadays built of thermal blockwork or stud-framed and insulated panels; note that all three of these inner skins need to be tied to the outer skins with a traditional criss-cross arrangement of metal wall-ties.

Inner string: The long board-on-edge that houses all of the step-ends of a staircase at the closed side (against the wall) of a typical wooden flight of stairs. Note that the inner string is more commonly referred to as the *wall string*.

Insect infestation: See **Woodworm** and/or **Death-watch beetles**

In-situ concrete: Concrete structures formed in their actual, intended location, such as reinforced-concrete floors – as opposed to *pre-cast* reinforced-concrete floor-beams, that are factory-made and transported to the construction site.

Inspection chambers: See *Figures 10 and/or 52* on pages 21 and 142: In an underground drainage system, these chambers (historically referred to as *manholes*) are positioned at predetermined points in the grounds of a building to suit discharge-outlets from SVPs (soil and vent pipes) and from RWPs (rainwater pipes) and waste pipes, discharging into gullies. The other consideration is that any obtuse or right-angled bends in a drainage run – to suit its exit route to the public sewer in the road – must be exposed by open channels in an accommodating chamber.

Traditionally, the chambers were built with concrete bases of at least 100mm/4 in. thickness, surrounding brick walls of 225mm/9 in. thickness, that were capped with cast-iron covers. The SGW (salt-glazed earthenware) segmental-shaped channels therein (connected to and from the inlet and outlet pipes), were built-up at their sides with sloping and partly-rounded, smooth-surfaced sand-and-cement *benching*. Its pur-pose being to eliminate any flat surfaces that might retain discharged faeces and tissue paper, etc.

Modern-day inspection chambers are usually of a circular shape of various diameters and are made of polypropylene, with numerous fixed or adjustable inlets/outlets that connect to 110mm/4⅜ in. diameter uPVC underground drainage pipes.

Insulation to flat roofs: See *Figures 18 and/or 21* on pages 46 and 55: The Building Regulations' *Approved Document F2* states that excessive condensation in roof voids over insulated ceilings must be prevented, otherwise the thermal insulation will be affected and there will be an increased risk of fungal attack to the roof structure. This applies only to roofs where the insulation material is at ceiling level (cold deck roofs). Where the insulation is kept out of the roof void, placed on the deck (warm deck roofs), the risks of excessive condensation developing are not present and these roofs, therefore, are not mentioned.

Intercepting traps: *Figure 52*: These are usually found to have been used in brick-built manholes (now referred to as *inspection chambers*) serving old underground-drainage systems that used 100mm/4 in. diameter, glazed-earthenware pipes. The final, out-flowing chamber of a site drainage system usually contains a glazed earthenware intercepting trap, which is jointed to the chamber's half-round, open-channel and discharges to the public sewer via a so-called *p-trap* – similar to that used for WCs – to inhibit unpleasant odours rising up from the sewers. The intercepting trap also usually has an integrated branched-pipe at its top, which only shows in the chamber as a plugged outlet, just above the mouth of the trap. This plugged – or unplugged – outlet is a so-called 'rodding eye', to be used for clearing any blockages. If the plug is found to be missing – as they often are – a replacement should be recommended to be fitted to inhibit noxious methane gases from rising up from the sewers.

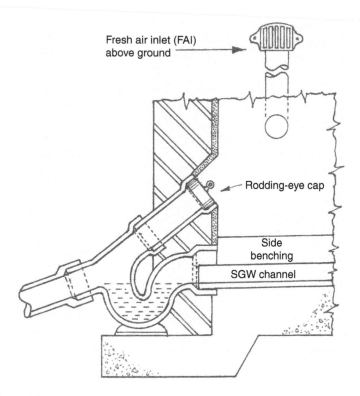

Figure 52 Intercepting trap in a traditional inspection chamber

Interlocking tiles: See *Figure 30* on page 69: These ubiquitous concrete tiles – technically known as *single-lap tiles* – are produced in a number of top-surface patterns; of which the Redland 49 seems to be the most popular nowadays. They are interlocking via their 25mm/1 in.-wide corrugated side-laps (and such *weathered* side-laps – of half-tile thickness – allows them to be laid with single-laps, as opposed to double-lap *plain tiles* which require a greater number of tiles). Hence the popularity of interlocking tiles, i.e., less tiles equalling less cost and less labour. Note that the required *lap* for these tiles should be between 63mm/2½ in. and 75mm/3 in. The term *lap* refers to the distance that the bottom edges of the tiles are required to extend over the heads of the tiles already laid below them.

Interstitial condensation: Condensation trapped in-between differing materials, such as between plasterboard and thermal insulation in an unventilated flat roof.

Intrados: In geometrical terminology, this term refers to the curved underside surface of an arch.

Intumescent paint: Such paints, that swell up to form a foam barrier, are used nowadays to aid fire-resistance. For example, if a door lock or latch are to be mortised into the edge of a fire-resisting (FR) door, intumescent paint might be specified to be applied to the interior surfaces of the mortise-hole cut in the door.

Intumescent strip: Different grades of intumescent strips are required around certain specified door-edges nowadays, which must be manufactured to BS (British Standards) fire-safety numbers – these are specified to cover *FIRE and SMOKE* situations, or *FIRE ONLY*.

Invert-depth: See *Figure 22* on page 65: Although seemingly contradictory to a dictionary's definition of *invert*, in building-terminology, *invert* refers to a depth-measurement below a theoretical datum line (known as the *collimation height*) to the centre of the *inverted*, upper surface of an open, concaved (semi-circular-shaped) drain-channel. Traditionally, this channel will be of salt-glazed earthenware (SGW), but in recent decades it will be found to be of uPVC underground-drainage quality. Note that invert-depths of an underground-drainage system are set up between the number of inspection chambers to achieve the required *falls* (gradients) of the pipework en route to the public sewer. Also, it is more likely on new-build sites that the inspection chambers will be of built-up circular sections of polypropylene-formed units. The bases of which, have pre-formed side-inlets to receive the various drainage pipes – angled into a self-contained semi-circular drain-channel.

Ironmongery: See **Door furniture**

Ironwork: A general reference to wrought- or cast-iron items, such as railings.

IWSc: The Institute of Wood Science, now amalgamated with *IMMM* (The Institute of Materials, Minerals and Mining), who recently changed their initials to *IM3*.

J

Jack rafters: See *Figure 2* on page 6: These are short rafters with a double (compound) splay-cut at the head (to fit against a hip rafter) and a birdsmouth-cut near the bottom (to fit onto a wall-plate's edge), which are fixed in diminishing pairs on each side of a hip rafter.

Jamb(s):   This term refers to the side-jambs of door- or window-frames. But historically, the term referred to the side-*walls* of door- or window-openings; which are now referred to as *reveals*.

Japanese knotweed: This bamboo-looking plant, which was apparently brought into the UK in the mid-1850s and was popular with landscape-gardeners, has now become a big problem in this country. By the early 1900s, people were being warned to keep it in check. It became known that its quick growth also spread quickly through its root system, causing thick clumps to spring up in nearby areas up to heights in excess of 2m/6 ft. 6¾ in. Apart from the ability to spread into neighbouring gardens, its root system has historically attacked and damaged aged drains and inspection chambers (via existing thermal- or settlement-cracks), boundary and retaining walls, paths, patios, driveways and outbuildings, etc. Its presence – or even its documented treatment of eradication – can also jeopardize the sale of a property and create mortgage and insurance problems.

For initial recognition, the developed leaves are a vibrant green colour, with a bloated heart-shape and a flattish base – which have been very well-likened by a RICS' Information Sheet (IP 27/2012) as being *shield*-shaped. They alternate on each side of a slightly zig-zag stem.

Jerkin-head or **hipped-gablet roof:** See *Figure 49* on page 122.

Jetty: The traditional projection at first-floor-level of an elevational wall (or walls) above the façade of the wall below. This was achieved historically by *jettied* (projecting) floor-joists (beams) supporting bressummer beams above. For more details of this ancient practice, see **Tudor and mock-Tudor façades** herein.

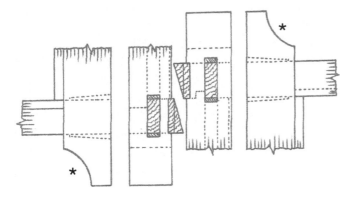

Figure 53a *Joggles (horns) below and above sliding sashes

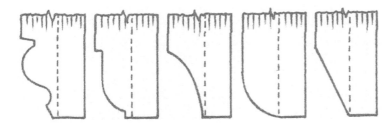

Figure 53b Typical sash-joggle shapes ranging from traditional ogee to modern splay

Joggles: See *Figure 53(a)+53(b)*: These are ornately-shaped sash-stile extensions (horns), that project past the outer edges of a sliding-sash's meeting rails. They enable the meeting-rail joints to be mortised and tenoned, instead of being dovetailed. The latter – which was done in the Georgian period of 1714 to 1830 – being more time-consuming and costly.

Joinery: The making and/or manufacture of building components, such as stairs, wooden windows, cupboards, doors, etc., that require to be fixed on-site.

Joint(s): **1)** A reference to the horizontal and/or vertical *mortar joints* of brickwork or blockwork, etc., or **2)** a reference to the tolerance gaps (*joints*) of 2mm to 3mm around doors and windows, to allow for thermal- and opening-movements, or **3)** a reference to the close- abutment joints of a built-component's framework.

Joist hangers: *Figure 54*: Steel joist-hangers are manufactured from 1mm-thick galvanized steel with pre-punched nail holes. The two main types are referred to as TT (timber to timber), of which there are *long* and *short* versions, identified as TTL and TTS, for differing depths of joists – and the other type is referred to as TW (timber

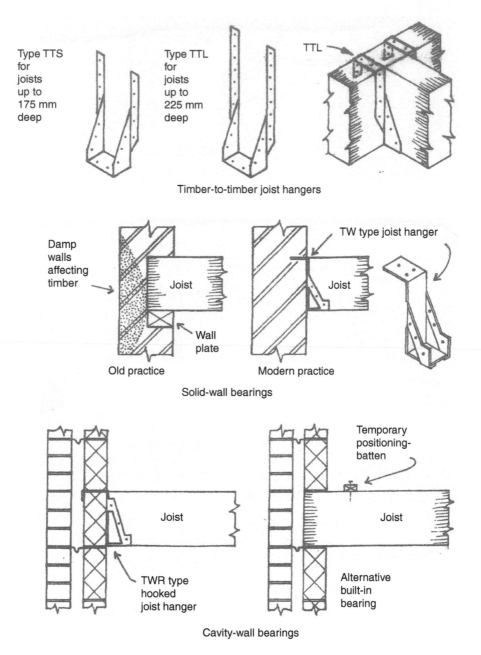

Type TTS for joists up to 175 mm deep

Type TTL for joists up to 225 mm deep

TTL

Timber-to-timber joist hangers

Damp walls affecting timber

Joist

Wall plate

Old practice

TW type joist hanger

Joist

Modern practice

Solid-wall bearings

Joist

TWR type hooked joist hanger

Temporary positioning-batten

Joist

Alternative built-in bearing

Cavity-wall bearings

Figure 54 Galvanized-steel joist hangers

to wall). The TT hangers are for the fixing of end-grain trimmer-joist abutments to the faces of trimming joists, wherever an opening is required in a joisted floor. The fixing of the hangers should be done with 32mm/1¼ in. galvanized nails. The TW (crank-shaped) joist hangers, made from 2.5mm-thick galvanized steel, are for end-of-joist wall bearings. Note that owing to a double metal-flange at their base – of 5mm thickness – the bottom edges of the joists require notching to achieve a flat surface for the plasterboard ceiling.

Joists: The skeletal ribs of a floor, traditionally in the form of solid timber joists suspended on their edges between wall-bearings or joist-hangers – or in a modern form of engineered joists – produced in factories. See **Engineered joists**.

Judas' hole: A peep hole in a door, usually in the form of a small diameter, magnifying-glass fitting, but also as a very small, inward-opening hinged-panel at average face-height level.

Junction boxes: Used by electricians to create designated socket outlets in a ring-main circuit – or for breaking into existing ring mains' circuits to extend the ring for the purpose of adding more socket outlets.

K

Kerf: A shallow or deep sawcut, of equal depth, across the grain of timber.

Kerfing: A method used to bend a timber component by making a series of sawcuts (kerfs) across the grain, thereby enabling it to be bent to a concaved shape on that side. The smaller the radius of the curve, the closer the cuts. Traditionally, this was done to the face-sides of small-sectioned skirting boards made to fit around the apron wall below a segmental-shaped bay window.

Keystone: The uppermost, crown-positioned *voussoir* (taper-shaped brick or stone) of an arch. Its final insertion *lock*s the arch structurally. Hence the name *key*stone.

Kicking plate: A thin metal plate, of bronze, aluminium or brass, etc., usually about 200mm/8 in. deep, screwed across the base of much-used doors in offices, etc., to protect the door's finished surface from accidental kicking.

Kiln-seasoning of timber: Once trees for commercial use have been felled, then cut into transportable boles (logs) and eventually sawn into basic sectional-sizes, the process of reducing their moisture-content is carried out. This is essential, because unseasoned (green) timber holds too much moisture in its cellular structure to be used commercially. Nowadays, the pre-sliced logs (separated by *piling sticks*) are stacked in large kilns (ovens) fitted with heaters, steam-pipes and fans – in which a carefully controlled programme of regulated heat, humidity and air circulation is carried out. Essentially, kiln-seasoned timber can achieve the required moisture content of below 20% mc., which is required in centrally-heated property.

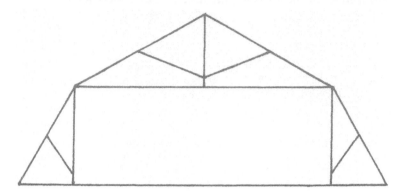

Figure 55 Line diagram of a King-post roof truss superimposed on a Queen-post roof truss (thereby creating a Mansard roof)

King-post roof truss: See *Figure 55*: A historic timber truss (still seen in large, well-built roofs) consisting of a horizontal tie beam, a central kingpost that rose up from the beam to support the ridge board, and two side struts that emanated from the lower sides of the kingpost – forming a Y shape – to support the principal rafters and purlins on each side.

Kite winder: On traditional tapered (winding) stairs, that used three tapered treads to affect a 90⁰ (quarter) turn, the central tapered-tread was called a *kite* winder, because its horizontal appearance (in plan-view) gave the impression of a kite. The tapered tread above that was referred to as a *skew* winder and the tapered tread below the *kite* was called a *square* winder. Note that all three terms related to the angles created between the stair-strings and the steps' front nosing-lines.

Knocking up: This trade terminology has two meanings: **1)** *Mixing* a quantity of mortar or concrete, etc., for immediate use, or **2)** *Remixing* the same material after it has exceeded its initial setting-stage – and (with more water added) it is *knocked up* to make it pliable enough to use. This is bad practice, which weakens the remixed material structurally.

Knotting: A shellac used for sealing knots prior to applying the priming coat, to stop them *grinning* or *bleeding* through the layers of paint and/or exuding resin. When knotting has not been applied, brown knot-shaped stains usually appear on the finished paint-film surface, spoiling the visual finish – and indicating an amateurish paint-job. Shellac is derived from an incrustation formed by lac insects on trees in India and nearby regions.

Knuckle: The middle, pivoting area of a hinge, which interlocks like a comb-joint, through which the pin is fitted.

L

LABC (Local Authority Building Control) Warranty: See **Warranties on newly-built houses**

Laced valley: Such a valley is/was formed entirely with one-and-a-half-width plain tiles (usually referred to as *tile-and-a-half* tiles) – which are normally used at verge-edges to create the staggered plain-tiling bond. The inverted dihedral angles of the raftered valleys were built-up with valley boards – and the 1½-width tiles were laid and fixed in a diamond-shaped appearance up the valleys to abut the edges of the plain roof-tiles on either side of them. The *lacing* term seems well chosen, as the appearance of such valleys looks very much like a laced-up shoe, with the unavoidable wedge-shaped gaps between the crossed laces; as with the tiles.

Lagging: **1)** This is a traditional reference to closely-spaced narrow strips of timber nailed across the dual rib-structures of wooden *arch centres*. The latter were/are made to support geometrically shaped brick- or stone-arches whilst being built. **2)** This term is used in reference to the thermal insulation of pipes and tanks, etc.

Laitance: The milky-fat brought to the surface of unset mortar or concrete, etc., if it is tamped or trowelled. However, in *pointing* or *repointing* of mortar-joints, there is a critical timing element related to the degree of setting-time required before *trowel-pointing* or *concaved-pointing/ironing* can take place. Best results are achieved when the mortar is not too hard and not too soft. Ironing or trowelling (the former being the predominant technique nowadays) at the right time brings a degree of laitance and smoothness to the surface that improves its weathering lifespan. But over-ironing or over-pointed trowelling should be avoided, as this creates a laitance-skin that tends to break away after a few years.

Laminated plastic: This refers to the relatively modern process of soaking many layers of craft paper or textiles, etc., with synthetic resin and compressing them at a very high temperature to form sheets of strong, durable, hard-wearing material. Such plastic laminates, with Registered names such as Formica and Wareite, etc., have been used for worktop-type surfaces, etc., for many decades.

Laminate flooring: This generic term refers to *overlay flooring* which is superimposed on top of subfloors. They may be of laminated timber, with a hardwood top-veneer, or laminated plastic, often with a very realistic impression of a variety of hardwoods on their top surface – even with minute woodgrain indentations.

Hardwood floors and overlay floors, however, are not new. Wide, square-edged oak floorboards (with widths up to 300mm/12 in.) were used centuries ago in a number of stately homes – and hardwood-strip floors (comprised of narrow, ex 75mm/3 in. strips of tongue-and -grooved boards) have been used for overlaying subfloors for many decades. Traditionally, they were *secretly nailed* through the inner edges of the tongues, into a softwood-boarded subfloor – whereas nowadays they are more likely (via PVA adhesives) to be glued together and laid as a *floating floor*. However, relatively thin, plastic-laminate floors – with no option other than being laid to float – use very clever self-locking click joints without glue. And they are held down (at their essential, expansion-jointed perimeter edges) by the provision of skirting. Where rooms are already skirted, it is more professional to remove them and refix or renew them upon completion – thus allowing the expansion gap of approx. 12mm/½ in. to be covered up. Alternatively, patent or quadrant beading can be fixed to the in-situ skirtings.

However, if the existing skirting boards are a fair depth – say at least ex 150mm/6 in., minus its top moulded-edge – it should be possible to superimpose another, lower-height skirting board (say, of ex 75mm/3 in. height) on its lower face. This would cover up the expansion gap and hold the floor down. It will look rather like traditional, step-faced skirting, but – to my mind – it would be a better alternative than amateurish-looking quadrant-shapes. Of course, at architrave abutments, the lower portions of these would require cutting off and being replaced with plinth blocks (see *Figure 5*).

Laminboard: This refers to a sheet material which is very similar to blockboard, but is far superior. Like blockboard, it is available in sizes of at least 2.4m x 1.2m/8ft. x 4 ft., with common thicknesses of 18mm/¾ in. and 25mm/1 in. The superiority of this board is gained by the thin core strips of timber (that run lengthwise through the boards) being only 7mm/¼ in. wide in their side-by-side build-up across the board's width. This minimizes tangential shrinkage and ensures a much flatter surface (less visibly undulating) on the face plies.

Landfill: In building terms, this is a reference to areas of undulating land that has been made-up historically (for some known or unknown reason), or has been made-up/filled to achieve a level site. The important point is that houses/buildings should not be built too near a landfill site. Such sites may be recorded by Ordnance Survey as *infilled land* and give information on a colour-scheme map, showing the nearness of a previous pond, quarry, mine, etc., that has been landfilled. Such made-up ground is filled with various materials, which can cause structural problems to properties built too close to them. Nowadays, the proximity of landfill sites is usually highlighted after a solicitor's initial environmental search-procedures, via specialist companies such as groundsure.com, whom they contact on their client's behalf.

Landing: This term usually refers to a stair-landing, which may be at the beginning of a stair, in the mid-area or at the top. If in mid-flight, it may be thought of as a brief resting place (for an elderly or unwell person), but mid-area landings are basically used to effect a change in the direction of flight. In this respect, they are therefore referred to as quarter-turn or half-turn landings – i.e., the *quarter* changes the direction of flight by 90^0, the *half-turn* changes the direction by 180^0.

Landslip: As the term suggests, this refers to the sliding down of an area of soil or rock – usually from a steep slope, such as a cliff.

Lantern light: This roof light, in its traditional form, looks like a miniature conservatory. Although, it seems certain that its name likens it to an actual lantern-light that could be carried around or hung by the metal ring at the apex of its pyramid-shaped glazed roof, above its four glazed sides. Such is the formation of a traditional lantern light, built on a flat roof. But they are usually an oblong shape and their raised, glazed sides support a double hip-ended pitched, glazed roof. The sills of such roof lights are fixed to a raised curb that is an integral part of the weathered flat roof. The side or top windows may also be openable to provide ventilation.

Laser scanning: During a site- or building-survey, a laser scanner can be used to obtain a photographic record of an area or object, via visualization and modelling software. This produces a three-dimensional image via the creation of 'point clouds' of data from the object's surfaces. Such surveying methods are much quicker than previous practices that used traditional surveying equipment and methods of measurement.

Lateral-penetrating damp: This surveying term refers to numerous situations that causes moisture to pass through a structural wall and dampen the interior surface. Such incidences may be explained as follows:

1)   Traditional single-leaf (skin) brick walls of 225mm/9" (one brick-length thickness), were theoretically reckoned to withstand damp-penetration from driven-rain for up to half of their thickness, on the assumption that during non-rainy periods the walls would dry out and not affect the inner half of their thickness. But this did not take into account the usually poor maintenance of exterior wall-surfaces and the breaking down (erosion) of the protective *pointing* of the mortar joints. Notwithstanding that – in the 19th century – lime (which is very hygroscopic) would have been used in the bedding-mortar. So, with defective pointing and weak mortar, lateral-penetrating damp can be drawn across the full thickness of such single-leaf walls, to the interior plaster surfaces, via capillary-attraction (capillarity); Note also that such lateral dampness can affect any timber joist-ends – and wall-plates, carrying the floor joists.

2)   Overflowing gutters or defective hopper-heads or downpipes; the omission of cavity trays; the bridging of cavity walls via instances of built-up mortar-droppings onto the metal wall-ties are other instances that can cause lateral-penetrating damp;

3)   Taking into account the weathering theory expressed at **1)** above, whereby the inner half-thickness of a single-leaf, brick wall of 225mm/9 in. thickness, was thought to be moisture-free – one can see the logic of moving that inner half-portion away from the outer portion to create a cavity. However, this concept has become blurred in the last five decades, via the introduction of cavity-wall insulation, which is now being claimed to cause lateral-damp penetration in an unknown number of dwellings – and occupiers are paying specialist companies to extract it. This is covered herein under the heading of: **Cavity wall-fill extraction**.

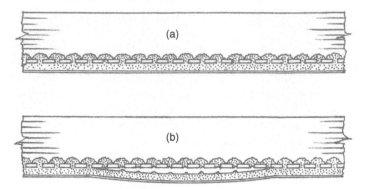

Figure 56 (a) Lath-and-plaster ceiling (fixed to ceiling joists). (b) Partial collapse of an area of ceiling, via loss of key.

Lath-and-plaster ceilings: *Figure 56*:   Such ceilings were formed in dwellings and other buildings up until the mid-20th century, until they were superseded by plasterboard. They still exist in a great number of properties – and consist of thin, wooden laths (of about 6mm to 7mm thickness x 25mm width x 900mm length). They were fixed to the underside of the ceiling joists with gaps between them of about 9mm/⅜ in. But because of their short length (900mm/ 3ft), the laths were fixed in the form of staggered, oblong panels, similar to the bonding pattern used on modern-day plasterboard-ceilings. This was done to minimize the length of end-butted joints. A so-called *pricking coat* of bovine-haired lime-mortar was then applied by a *laying-on trowel* (otherwise known as a *skimming trowel*) and it was squeezed through the gaps to form longitudinal mushroom-shapes on the topsides of the laths. These bulbous-shapes acted as 'keys' to 'lock' the mortar to the laths. When set, this initial rough surface was *rendered* with lime-mortar and trued up with a straightedge and a wooden float, before being skimmed (*set*) with finishing plaster. Such ceilings have stood the test of time, but, with age, they are prone to partial collapse through a loss of bonding 'key' between the laths and the main body of plaster; i.e., the longitudinal mushroom shapes – that hold up the plaster – tend to break off over time. Partial collapse, although unlikely to be life-threatening, is not uncommon. Bulges and cracks, therefore, should be considered carefully by surveyors on their inspections, if only for Health & Safety reasons and/or litigation. An incidence of a partial collapse (of about 900mm/3 ft. diameter) above a bed is known to me, but fortunately for the occupant it did not happen during bed-occupancy hours.

Lath-and-plaster used externally: When surveying/appraising early-19th-century dwellings, it should be borne in mind that the eaves' soffits will either be of painted softwood or painted lath-and-plaster – both of which are likely to be deteriorated. The former with wet-rot decay (perhaps mostly secreted by the paint film), and the latter with partial loss-of-key between the bovine-haired lime mortar and the wooden laths. With this in mind, as-close-as-possible examination via binoculars is essential.

Other lofty areas that require binocular-examination on this age of property, are dormer windows that have gable-ended pitched roofs above them. It was quite common for the infilled, triangular gable-ends above these windows, to be timber studded and clad with lath-and-plaster on the inner and outer surfaces.

Lath-and-plaster walls: Traditionally, timber-stud partitions were lined with lath-and-plaster, as described above for ceilings – and as were, to some extent, the inner surfaces of *exterior single-skin* (225mm/9 in. thick) *brick walls*. Having come across the latter on a number of surveys, it seems to me that they were an early form of *Victorian dry-lining*, to inhibit lateral, penetrating damp. Sawn timber-battens, of 50mm x 25mm/2 in. x 1 in. section, were nailed vertically to the walls at approx. 400mm/16 in. centres – and the arrangement of gapped laths was fixed to these horizontally, prior to the 3-coat lime-mortar plastering procedure. However, the untreated softwood battens fixed to the single-skin walls, are usually found to have suffered from wet-rot decay. But not as badly as the brick-sized softwood inserts in the wall – to which the battens are occasionally found to be fixed. No doubt such inserts (resembling offcuts of 100 x 63mm/4 in. x 2½ in. sawn wall-plate timber) had been pre-positioned when the walls were being built – in readiness for the batten-fixing and the Victorian dry-lining.

Lavatory: A Latin-derived word meaning a water closet (WC).

Lay bars: See *Figure 17* on page 42: This is a term used for glazing bars in a horizontal position in windows or doors. Note that lay bars in multi-glazed doors should be continuous (i.e., mortised to receive the half-depth tenons of vertical bars), thereby enabling them to better support the glass bearing on their slender outer-edges – but this rule should be transposed for lay bars in multi-glazed, sliding sash-windows, because the slim, horizontal meeting-rails of the top sashes, need the additional support from any mid-area mortise-and-tenon joints provided by vertical glazing bars.

Leaded-light windows: These are comprised of narrow strips of H-shaped lead that act as glazing bars to separate and retain small panes of glass embedded in putty – and these *glazing bars* are called *cames*, pronounced *cams* (the singular term being *came*, pronounced *cam*). Note that the traditional coloured putty – used as a bedding sealant within the cames – tends to deteriorate with age, sometimes causing rainwater to penetrate through the leaded light by capillary attraction. In such instances, re-glazing of the leaded lights would be advisable. Traditionally, leaded lights were reinforced with small-diameter steel bars wired superficially to the inner faces of the cames, but nowadays (if genuine leaded lights are replicated), the steel bars are secreted within the confines of the H-shaped cames.

Lead flat-roofs: Although lead is the most expensive covering for flat roofs, good quality lead roofs, laid with well-formed rolls and good detailed features, are reckoned to have a likely life-span of up to at least 100 years. The BS code number and colour code for thicknesses of sheet lead for flat roofs are 4 (blue) or 5 (red) for small roofs, dormers and their cheeks, flashings and valley linings – and 5 (red), 6 (black) or 7 (white) for large roofs, valleys and parapet gutters.

Lead nails: This puzzling term refers to nails made of copper used for fixing lead in certain places.

Lead paint: Such paint, which contained *red-* or *white-lead* poisonous pigments, stopped being made at least six decades ago. And although this is assumed to be common knowledge, it may not be known that great care should still be taken when doing renovation or redecoration work on older-built properties. If it involves removal (burning or scraping off), or hand- or mechanical-sanding of the old paintwork – it could still be very dangerous to inhale the dust. So, for safety's sake, a good-quality face-mask should be worn.

Lead pipework: Regarding incoming lead water-main pipes, it seems to have been widely believed in the past that there was no threat to health because of the protective coating that forms in the pipes when the inner surfaces oxidize. But this protection is not now considered to be sufficient and replacement of the lead main is recommended – but not enforceable. However, because the repair of lead pipework is disallowed now-adays, and repairs via copper pipes and fittings are disallowed (because of electrolytic corrosion), repairs have to be carried out with plastic pipes and compression joints specially designed for this purpose. Electrolytic corrosion is a galvanic action that can occur by joining two different metals together.

Lead saddles: These are pieces of sheet lead that have been *dressed* or *lead-welded*, to form various-shaped weathering-saddles that fit over-and-under tiled or slated roof-junctions. One such three-way meeting point, is where the pitched roof of a dormer window's right-angled ridge-junction, is attached to the upper-area of the main pitched roof and thereby forms a three-way sloping juncture; this is where a lead saddle must be dressed *over* the ridge and tiles of the dormer – and under the tiles of the main, pitched roof.

Lead safes (for chimneys):   Although these are technically referred to as *lead safes*, they are more commonly known as DPCs (damp-proof courses) to chimney stacks. However, they are not often seen to have been used and there are two likely reasons for this. One is that they interrupt the building of the stack, once it has been built to the DPC level, just above the roof-line, causing a delay until it is made (with upturned back, sides and down-turned front edge) and fitted by a plumber or roofer. The other reason is that there is an understandable degree of concern about weakening a chimney's stability, by separating its higher structure from its base area with a *lead safe* acting as a DPC. Even if the lead tray is coated on the top and underside with bitumen, as recommended by The Lead Development Association, the question is will that provide as good an adhesion as the mortar-bonded brick-to-brick courses above? It is a design problem, because

DPCs are really needed to inhibit descending dampness into the roof timbers below. But this has to be weighed up against the likelihood of weakening the stack's stability in a really windy storm or in a coastal area.

Lead- or copper-slate: This term refers to a special form of protective *weathering sleeve* – usually handmade and of sheet lead or copper – that fits closely around a pipe (usually a large diameter soil-and-vent pipe) at its exit point through a roof. The circular sleeve is angled at its base to fit the pitch-angle of the roof and it is welded or soldered to the *lead-* or *copper-slate*. The slate (so-called because it replaces an actual slate or tile) is in the form of a square piece of lead or copper, the sides of which measure about 300mm/12 in. long. And its centre has an elliptical-shaped hole cut in it to receive the attachment of the circular sleeve. They are then both welded/soldered together, ready to be fitted over the protruding soil pipe. The actual 'slate', when it touches the tiled or slated roof, must be carefully *dressed* into the contours of the surrounding tiles or flattened to the slates. Importantly, it must be fitted *under* the tiles or slates at its top. To visualize the lead- or copper-slates, they can be likened to resembling upside-down academic mortarboards – with a hole in the tilted mortarboard to receive the soil pipe protruding through the pitched roof.

Lead soakers: See **Soakers**

Lead wedges: These are small pieces/strips of sheet lead, folded and beaten together several times (to increase their thickness) and used as driven-in wedges to hold dressed-lead flashings – and stepped-flashings – into the raked-out, horizontal brick-joints. Note that this operation should not be carried out on *green* (newish) brickwork, because of the likelihood of breaking the mortar-joint's adhesion to the bricks.

Leaf: This term is sometimes used to describe one part of paired constructions, i.e., one of a pair of doors or windows; or one half of a cavity (double-leaf) wall, etc.

Leaking (underground) drains: See **Subsoil erosion and subsidence damage**

Leaning chimney-stack phenomena: Although domestic chimney stacks appear to be quite solid structures above a roof-line, they are in fact mostly rectangular, hollow shells of half-brick-thick (112mm/4½ in.) walls formed around 225 x 225mm, 9 x 9 in. chimney flues. And, internally, they are subjected to dampness from flue-condensation – which, mixed with soot deposits on traditional flue-linings, can occasionally cause patches of black, deliquescence-staining to appear on the room-side of a plastered chimney breast. For more information on this separate subject, see **Deliquescent soot-ash staining to chimney-breast plaster**.

Externally, these independent, hollow structures – if observed habitually over a period of time – can quite frequently be seen to be leaning over to the south, usually with a distinct, perceptible curve. This is scientifically due to sulphate crystallization and expansion of the rain-soaked mortar joints on the *north-side* of a chimney's brickwork, being more deprived of the drying benefits of sun and wind, than the southern aspect.

Lean-to roofs: These are usually used on parts of a building that extend beyond the main structure, comprised of mono-pitched rafters *leaning* on the structural wall in various ways, i.e., birdsmouthed onto a wall plate supported by wrought-iron corbels or supported by continuous brick-corbelling projecting from the wall's face. If the potential thrust of a lean-to roof can be discounted, the top plumb-cuts of the rafters may simply be fixed to a ridge board fixed to the wall. Modern-day lean-to roofs are usually formed with factory-made, mono-pitched trussed-rafter assemblies, which removes any potential thrust on the *leaned-on* structural wall. Note that a great number of yesteryear's lean-to roofs (and a lesser number of recently-built lean-to roofs) – if built onto a cavity wall – should be suspected of lacking a now-required cavity-wall tray; such being needed to inhibit dampness from the outer-wall above, soaking down to the inner-wall below.

Leasehold property: Legally, this refers to a property (or the land upon which the property is built) being leased (on loan) to the leaseholder for a stipulated period of years. Nearing the end of this period of time, the renewal of the lease has to be negotiated with the freeholder, via a solicitor, and an agreed sum of money paid.

Ledged and braced (L&B) doors: This refers to a traditional type of doors still in use. Originally, the tongue-and-grooved edges of the boarded faces were either bead-moulded or lambs-tongue moulded, but now they are simply formed with vee-jointed matchboarding. Also, the boards were reliably *clench-nailed* with 50mm/2 in. *black iron, cut clasp nails*, onto the two or three horizontal ledgers (which usually had sloping/ weathered) top edges. But nowadays they are usually screwed, which is not as strong. The first of two types, seen vary rarely nowadays, is comprised of square-edged or tongue and grooved (T&G) boards and three cross-rails, but *no* diagonal bracing, which made them prone to sagging. The second type, still widely used, is comprised of tongue and grooved matchboarding, near-top, near-bottom and mid-area ledgers and one or two diagonal braces that form either a single or a double Z formation, one above the other. These are ideally placed at 60^0 (if only one diagonal brace is used). When hung (on cross-garnet hinges) the positioning of the braces should act as gallows' brackets against the hinged side, to inhibit this door-type's tendency to sag. But, on surveys, one notices occasionally that this sensible rule has been ignored – usually with the result of the door having dropped significantly on the latched, opening-side.

LED lighting: These initials stand for *light-emitting diode* and such products produce up to 90% more light-efficiency than traditional incandescent light bulbs. LED lighting operates via a microchip, which illuminates the miniscule light-emitting diodes.

Letter plate: This ubiquitous piece of ironmongery, often referred to as a letter *box*, was traditionally fitted in the mid-area lock rail (which it still often is), but nowadays it is also found to be fitted in the bottom rail. This change is related to a greater variety of modern door-deigns, often taking over the mid-area of the door, disregarding the ergo-nomics of easy post-delivery. Although, perhaps more disturbing to the postal service, is the inner spring-loaded and draught-proofed backflap.

Levelling terms used in site-levelling: See *Figure 22* on page 56: There are a number of tripod-mounted levelling instruments available nowadays, but the common principle of their use is that they provide a theoretical horizontal line (plane), encompassing the whole building site when the head of the level is pivoted around a 360^0 circle. In levelling terms, this *theoretical plane* is referred to as the *height of collimation*, or the *collimation-height*, from which a variety of levels can be measured and/or established. These levels are taken (or established) from a measuring rod known as a levelling staff. On large building sites, the terms used in levelling are:

Collimation height: The height of a level's theoretical viewing plane above the ori-ginal Ordnance datum established at Newlyn in Cornwall.

Backsight (B.S.): The first staff-reading taken from the collimation-height above the site datum/OBM (Ordnance Bench Mark).

Datum: A solid and reliable fixed point of initial reference, such as an OBM or a nearby manhole (inspection chamber) cover in the road.

Foresight (F.S.): The last staff-reading before a levelling instrument is repositioned.

Intermediate sight (I.S.): All readings taken other than Backsight and Foresight.

Ordnance Bench Mark (OBM): A reference point (above ground), set originally by the National Ordnance Survey Department, as being a measured height above the mean sea-level at Newlyn in Cornwall, U.K., at a certain time and date.

Temporary Bench Mark (TBM): This refers to a temporary datum transferred from the OBM and set up on site at a reduced level – or to the same level as any alternative datum used, such as a manhole (inspection chamber) cover in the road – from which the various site levels are set up more conveniently. The on-site datum is usually in the form of a *datum peg*, of either a 19mm/¾ in. diam-eter steel bar, or a 50 x 50mm/2 x 2 in. wooden stake, set in the ground, usually partly-encased in concrete.

Reduced Level (R.L.): Any calculated level position above the original Ordnance datum.

Light-tunnels: See **Skylight sun-tunnels**

Lime mortar: This traditional bricklaying mortar consisted of lime and sand mixed with water to a ratio of 3 parts (portions) of sharp sand to 1 part of lime.

Linenfold-carved and moulded panels: These were used in stately homes and other such buildings built in the 15th/16th century and such door panels and wall-panelling can still be seen occasionally (or fleetingly glimpsed in scenes from *Downton Abbey*), displaying their extravagant linenfold-moulded and hand-carved scroll-patterns at the top and bottom areas of the folded-linen contours.

Lining: In carpentry terms, this usually refers to a *door lining*. It differs from a *doorframe* in as much as the latter is more solid. A door lining has relatively thin side-*legs* (which are the same thickness as the lining's *head*) of usually 28mm/1⅛ in. finished size. The joints between the legs and the head (on good-class work) should be narrow-tongued housings. But, regardless of any class-distinction, tongued-housings are rarely seen nowadays – such being another victim of overspent time equalling less money earned.

Lining paper: Traditionally, in good-class decorator's work, the plastered walls were lined with thick, plain *lining paper*, prior to wallpapering – to soften the appearance of any small *bumps* or *hollows* in the wall. This was also done to ceilings, particularly if the ceilings were of lath-and-plaster construction – and were displaying typical signs of cracks caused by *loss-of-key* (plaster detachment from the wooden laths). When surveying such ceilings, experience can lead one to believe that, in some cases, it is only the cross-layers of lining paper that are holding up the ceiling.

Link-detached properties: See **Detached** and **link-detached houses**

Lintels: See *Figure 57* on page 166: Metal, reinforced-concrete, stone or wooden beams over door- or window-openings, to support the structural loads above.

Lip/lipping: This is the term used for thin (usually 9mm/ ⅜ in.) strips of wood that *lip* the edges of birch-faced blockboard, laminboard and plywood, to cover up the core-structure of such material – and, if the lipped boards are to have a stained and/or clear varnished finish, Ramin or beech, two light-yellow coloured hardwoods, are good for this purpose.

Listed properties: See **Grade I and Grade II listed buildings**

Litigation threats to surveyors: Regardless of what type of surveying-inspection is being carried out (as described under **Building Inspections** herein), surveyors are responsible for *thorough* inspections involving fault-finding. Such faults – whether major or minor – must be appraised carefully and reported upon honestly. However, it must be borne in mind that it is not a surveyor's responsibility to describe his or her findings in such a way as to unduly influence the purchaser's decision to withdraw their offer of purchase. Such withdrawals are usually a shock to the vendor and can result in a surveyor being sued for causing the loss of their purchaser.

Live knots: Such knots, being the opposite of *dead knots*, are acceptable in most carpentry or joinery operations, providing they are not too large and their fibres are unbroken and ingrown acceptably within the grain of the wood structure.

Load-bearing walls: Apart from exterior walls, that bear such obvious loads as the roof and intermediate floors, etc., this reference usually refers to interior partition walls (of brickwork, blockwork or timber-studwork) that have been constructed to carry a load, such as timber ceiling-joists running at right-angles above them; and if the upper edges of the ceiling joists are within a roof-void, they are very likely carrying part of the roof's weight, via the thrust from purlin-struts. Identifying whether interior walls are load-bearing or cross-supporting requires a fair degree of acquired structural knowledge.

Local Authority: With reference to building, this refers to any local council's Building Control Department – which, in my area – has been changed in recent years to: The Building Control *Partnership*. This is still controlled by the local council, but the change is believed to reflect the subletting of their own building inspector's site-control to external self-employed building inspectors – who are no doubt technologically qualified.

Local Authority Building Control (LABC) approval: Apart from Planning Permission and Permitted Development, LABC approval seems to account for most of the activity involving legalized building work. And although owners can carry out repairs-and-maintenance work, there are not many jobs really that can be legally done without paying a council fee.

 However, to make the point, a few jobs requiring LABC approval are listed below: **1)** The installation of a shower, bath or WC., that involves new waste pipes or alterations to the drainage; **2)** The installation of a new heating appliance, i.e., a boiler; **(3)** Replacing roof-cladding (unless the replacement is exactly like-for-like); **(4)** Internal alterations, such as altering or removing a load-bearing cross-wall, or removing (partly or wholly) a chimney breast.

Location plan: These are sometimes shown on construction drawings, usually taken from Ordnance Survey maps, to identify the site and to locate the outline of the building in relation to the road(s) and its surroundings.

Loft: This reference to an isosceles-shaped roof-void (that traditionally housed problematic cold-water storage tanks) may also be referring to a habitable room formed in such an area. In such constructions, the ceilings of these rooms are very small, above which there is a small *apex roof-void*, which is usually inaccessible, via the omission of a trap hatch. This can be a problem, if such roof-voids require to be updated with insulation. See **Loft insulation**, below.

Loft conversions: There have always been obstacles to overcome regarding the design and feasibility of creating an additional room (or rooms) via a loft conversion in a dwelling – and these obstacles have now been exacerbated by more stringent requirements from the Building Regulations' Approved Documents (AD) being added, as follows: **1)** AD B1 + B2, refers to an early warning fire alarm system requiring to be added to the converted building; **2)** AD F refers to providing adequate means of ventilation; **3)** AD K1, K2 refers to stairs and ladders and *protection from falling* (i.e., balustrading); **4)** AD B1, B3 refers to floors, both old and new, requiring the full 30 minute standard of fire resistance – unless only one storey is being added, which contains no more than two rooms with a total floor area of less than 50m; **5)** AD F, Appendix E refers to required procedures concerning any existing fans and ducting in the loft; **6)** AD K1 refers to the regulations regarding the provision of stairs; and, finally: **7)** AD K3 refers to the protection from falling (i.e., balustrading).

Note that there are also references in AD B2, regarding provision for *emergency egress/means of escape* via a dormer- or roof-window (skylight), which should have an unobstructed, opening window of at least 0.33m² and be at least 450mm high x 450mm wide. The bottom window-*ledge* of such an escape-opening should be not-more-than 1.1m (approx. 3ft 7¼ in.) above the dwelling's floor level. Also, the bottom ledge of the escape window, in relation to the roof-slope below (before it meets the eaves), must not measure more than 1.7m (approx. 5ft 7in). This is to allow access and escape via a fireman's ladder.

The obstacles that have always existed regarding loft conversions, include: **1)** Is there enough headroom between the underside of the ridge board and the top of the ceiling joists? (A minimum of 2.3m/7ft 6½ in. is needed); **2)** Can a stairway/staircase be sited below, without losing/spoiling an existing room? **3)** Where will it end up in the loft-area, in relation to the required 2m or 1.9m headroom required above it? **4)** Do the existing, shallow-depth *ceiling* joists have enough cross-wall support below to act as floor-joists? Or, do they need to be increased in depth? **5)** Are there any service pipes, water tanks, boilers or extractor-ducts that may need to be repositioned? And, if so, does that seem feasible?

Loft insulation: Radical changes have been made to the Building Regulations in the last two decades, regarding *conservation of fuel and power*. These initially came into force in April 2002 as Approved Documents (AD) L1 and L2, but have now been amended and expanded into four separate documents, known as ADL1A, ADL1B, ADL2A and ADL2B. Regarding loft insulation, to achieve tabled U-values in the thermal elements, the insulation would need to be in the form of mineral-wool quilt, such as Crown Wool of 100mm/4 in. thickness, laid between the ceiling joists – with an additional 170mm/6¾ in. thickness laid over and across the ceiling joists. And to avoid cold/thermal-bridging, partial-fill or full-fill insulation in the tops of the surrounding cavity walls, should continue up to connect with the loft insulation.

However, where electric light fittings – such as recessed spotlights and downlighters – penetrate the insulated ceiling, they should be kept clear of the insulation by means of light-guard sleeves, or pre-formed pockets.

Loft ladders: Not found in a number of houses, especially pre-WWII properties, even though access to the lofts is usually provided via a lift-up or drop-down trapdoor. Loft ladders are usually of aluminium or stress-graded softwood and are either telescopic in two or three parts, or are of a fold-away type, that are integrated with a fully insulated trap door.

Loose-fill insulation: See **Cavity-wall problems**

Loose-pin hinges: These type of butt hinges have withdrawable pins that can fairly easily be withdrawn from the top of their knuckles. Should the need arise, this enables a door to be quickly removed without any unscrewing of the hinges.

Lost-head nails: These nails are technically known as *lost-head wire nails* because their barrel-shaped heads only protrude very slightly. For this reason, they are much used for fixing tongue-and-grooved floorboards and – if *punched-in*, as they should be – the usual moisture content in the boards can partly close up the pushed-aside wood fibres and make the nail-head-holes less noticeable; (hence the name *lost head*).

Louvres: These, in the form of horizontal slats in a variety of different materials, are spaced apart in an enclosed frame to provide ventilation. The angle of fixed slats can vary, but about 15^0 to the horizontal plane (at least) is necessary to keep out rain.

Low-level WC cistern: There are, in basic terms, three types of domestic WCs (water closets), related to the position of their flushing cisterns. First, is the WC with a *high-level cistern*, a very long flush-pipe and a pull-chain-flush operation; second, is the WC with a *low-level cistern*, a short flush-pipe and either a lever-flushing handle on the face of the cistern, or a modern duel-flush, push-button on the cistern's top; third, is the close-coupled WC, whereby the cistern sits on the WC and does not require a flush-pipe – and, again, has either a lever-flushing handle on the face of the cistern, or a modern dual-flush, push-button on the cistern's top. If interested, see **Water closet (WC)**, written herein, for more detail.

Lichgate or **Lychgate:**   An oak, churchyard gate, being at the entrance to a small, oak-framed, open-ended outbuilding with a pitched-roof, originally used to accommodate a *bier* (a wooden stand), upon which to rest a loaded coffin (*Leiche*, in German, means corpse).

M

Macerator-pumped WC-system: When an *additional* WC is required to be installed in a dwelling (or other type of building) and its only location does not allow direct access and entry into the property's established, vertical Soil & Vent Pipe (of 110mm/4⅜ in. diameter), that discharges its waste matter and foul water to the property's sewage system, then a macerator, with its smaller diameter piped-and-pumped system, may be permitted – after notification to (and approval from) the Local Authority Building Control department.

The macerator pump, in a uPVC box-like, sealed container, sits on the floor, immediately behind the WC pan and receives the pan's P-trap into the face of its upper area. When the WC's flushing lever is operated, the macerator pump comes into action, receives and breaks up the solid faeces as the effluent is pumped up vertically or horizontally and discharged into 19mm/¾ in. diameter pipes. Note that macerator-pumps can be quite noisy and are usually only guaranteed for five years.

Made-, make-, making-good: These different tenses of a traditional building term refer to a particular action of repairing or returning something disturbed to a good state of repair.

Made-up ground: See **Landfill**

Maintenance period: Relevant to new-build properties and/or major works, builders are usually contractually bound to a maintenance period of six months after completion or hand-over, in which to access and carry out maintenance of any defects that have occurred. These defects – which are usually caused by ubiquitous thermal movement – are, or should be, noted on a so-called *snagging list*.

Maisonnette: This French word means a small house or flat, but its use in the UK usually only refers to a self-contained flat within a dwelling house containing only one other flat. In the best sense of this definition, both flats should be completely self-contained, i.e., with their own separate entrances, without the need to share a communal hallway. However, this true definition has been blurred and there are flats around that are referred to as maisonnettes, by virtue of the fact that they occupy two floor-levels of the same dwelling – regardless of the important fact that they have shared communal access.

Maisonettes may be leasehold with a separate freeholder, or they may share the freehold with the owner/occupier of the other maisonnette. From a legal point of view, this second scenario is believed to create so-called *flying freeholds* – which means that each freeholder has to agree to sharing the cost of maintenance and repairs – which can, of course, be problematic.

Manholes: See **Inspection chambers**

Mansard roofs: See *Figure 49* on page 122: These are traditional roof-designs that are virtually comprised of one pitched roof superimposed upon another. They can be seen as either gable-ended or hip-ended. The lower, steeper roof-slopes, which were vertically studded on the inside (and usually tile-hung on their exterior), acted as walls and accommodated habitable rooms in the roof space. And the upper, shallower roof slopes had horizontal ceiling joists acting as ties, giving triangular support to the otherwise weak structure. Traditionally, these roofs usually incorporated king-post trusses superimposed on queen-post trusses. But nowadays, this design of roof is built using modern techniques, especially involving steel beams, taking their bearings from the gable-ended walls.

Mantlepiece/Mantleshelf: A shelf over a fireplace, usually attached to a fire surround.

Marbling: Using paint to create a realistic, decorative appearance of marble – or other types of veined stone – by a very skilled, traditional painting-technique.

Masking: **1)** In painting and decorating, this refers to the edges of finished surfaces – such as emulsion-painted walls, being carefully covered with *masking tape* up against the top edges of the undercoated skirting boards, to allow the finishing coat to be applied to the boards, without the skilful, but time-consuming, traditional task of *cutting-in* to the walls; **2)** In kitchen-fitting work, this term may refer to *masking tape* being carefully positioned on each side of an abutment joint between adjoining worktops, to limit the application of a coloured filler to the joint-area only.

Mason's mitre: In stonemasonry or featured brickwork, this describes a butt joint (usually right-angled) of two pieces of stone or brick, etc., that are chamfered or moulded – whereby the angle-mitred *appearance* is actually formed in the abutted stone or brick.

Mastic: This refers to a plastic, waterproof product, that only hardens on its surface, making it an ideal sealant for such uses as bedding window- and door-sills, sealing expansion joints and abutment-joints to window- and door-frames, etc.

Mastic asphalt/pitch asphalt: A manufactured mix of 95% crushed stone, gravel and sand, mixed with 5% coal-tar pitch, to make solid, tablet-shaped lumps (of about 250mm diameter). These tablets become fluid when melted down on site in an asphalter's pot and the semi-stiff asphalt is spread/laid as an impervious flat-roof membrane, usually in two layers. Additionally, where flat roofs abut walls, vertical, asphalt skirtings (or *plinths*) are formed to a minimal height of 150mm – and these skirtings are finished (as a separate operation) with 45° angled, asphalt fillets formed at their juncture with the asphalt roof-surfaces. The total thickness of finished asphalt is usually approx. 19mm/¾ in.

Matchboarding: This term refers to boards (commonly used for facing ledge-and-braced type doors) that are tongue-and-grooved on their opposite face-edges, of which one or two are moulded as well. The moulding is more commonly a so-called vee-joint (i.e., each edge is chamfered, making a vee-shape upon assembly). In one design of traditional matchboarding – both edges were also moulded, but with a so-called *lamb's tongue* moulded-shape. The other type was only moulded on one face-edge (the edge with the tongue protruding), with a bead-mould. Note that the lamb's tongue moulded matchboarding predates the bead-mould, which, in turn, predates the simple, present-day vee-joint.

Mat-recess: A sunken portion of floor – of about 25mm/1 in. to 40mm/1½ in. depth – close to an entrance door, to retain a piece of purpose-cut coconut matting.

Meeting rails: See *Figure 53(a)* on page 145: The two (usually centralized) horizontal rails of vertically-sliding sash-windows, that *meet* side-by-side when the windows are in a closed position.

Merulius lacrymans: See **Dry rot**

Metal lathing: See **Expanded metal lathing (EML)**

Metalled road: A road that is partly or extensively patched up or covered with broken stone, rubble, or with extensive patches of tarmacadam. Such roads are usually unadopted by the local authority.

Metal stud-partitions: Such walls are now erected on site in competition with timber-stud partitioning – and they are similarly clad with 12.5mm/½ in. thick sheets of plaster-board. Their attraction seems to be that the relatively cheap metal components are light in weight, more stable than timber studs (which are susceptible to shrinking, twisting and warping), easy to cut and fix – and are usually quicker and therefore cheaper to erect than timber studwork.

The metal studs and tracks are made from pre-galvanized, folded mild-steel of varying thicknesses ranging from 0.5mm to 0.9mm. The studs, of which there are two types, have preformed holes in their inner webs (or partly-formed holes, ready to be knocked out, if required) to accommodate service cables, etc., that might need to run through the partition.

The metal components for these partitions – because of their sectional shapes being likened to capital letters – are referred to as *U-tracks* for the *head- and floor-plates*, *H-* or *I-studs* for the more common *intermediate studs*, and *C-studs*. The latter are used as *wall-studs* and *door studs* – and as *end studs* when partitions are designed to change direction – thereby creating right-angled external angles (quoins).

Metal windows: See **Crittall casement windows and doors** and **Aluminium windows and doors**

Microbore pipes: Such small-diameter copper pipes were infrequently used tradition-ally in domestic central-heating systems, usually of 8mm external diameter.

Mid-feather: See **Wagtail**

Mid-withe: A traditional reference to an internal half-brick-thick wall division between two chimney-flues in a chimney stack.

Mineral-faced bituminous roofing-felt: This traditional material – not now in its original form, as it previously contained asbestos fibres – is used on flat roofs over garages and dormer windows, etc. It forms the top layer of a system referred to as *3-layered built-up roofing felt*. The two felt-underlays are of a lesser quality. The top layer is now comprised of a base material (the felt) made up of hessian or fibreglass and polyester, saturated with a coating of bitumen asphalt. Its textbook lifespan is reckoned to be about 10 years, but in some cases – with perhaps minor repairs – it has been known to double this figure.

Mist coat: This term in painting and decorating usually refers to the first coat of emulsion applied to a freshly plastered wall, whereby the emulsion is watered down by about 50%. This *seals* the plaster, which otherwise tends not to accept undiluted paint. However, other, patent sealers are also available nowadays.

Mitre: An angled cut (usually of 45⁰) made between two pieces of timber (or other material) of identical cross-sectional size and shape, as a means of forming a 90⁰ right-angled joint between moulded-edge components such as architraves and moulded-edge skirting boards.

Mitred knee: A traditional term that refers to a portion of geometrical, curved handrailing which was likened to a *bent knee*. Its position in the stair-balustrade occurs when the raking handrail has to rise more steeply to meet – and be mitred to – a level hand-rail above, on a landing. The mitred knee, therefore, is comprised of a concaved surface on its lower portion and a vertical surface below the mitre. Its lower joint is handrail-bolted and its upper mitre is reinforced by a secreted, sunken angle-bracket on its underside.

Mock-Tudor façades: See **Tudor** and **mock-Tudor façades**

Modern, designed floor-joists: See **Engineered joists**

Modern framing anchors: See *Figure 48* on page 120: Metal timber-connectors are now extensively used in first-fixing carpentry floor-joisting work, to replace traditional housed and-tusk tenoned framing-joints. These framing anchors are more commonly referred to as timber-to-timber joist-hangers. The advantages gained in their use is a saving in labour hours and more effective support of the trimmer or trimmed joists, by the bearing point being at the bottom of the floor's load. However, it must be mentioned that traditional housed and tusk-tenoned joist-framing joints have held up to the test of time in houses and buildings for several hundreds of years.

The most popular, widely used framing anchors are U-shaped *joist hangers* of 1mm galvanized steel, that are fixed to the side of trimming-joists to receive the ends of trimmers, or fixed to the sides of trimmer-joists to receive the ends of trimmed joists. The extended, pliable tops of their U-shape are bent and hammer-dressed over the recipient joists and fixed with 32mm/1¼ in. galvanized wire nails.

For joist-to-wall bearings, see **Joist hangers** written herein under that heading.

Modern lintels: *Figure 57*: Nowadays, these are usually of preformed, light-gauge galvanized steel, in a number of cross-sectional shapes to suit solid- or cavity-walls. The popular design for the latter, seems to be in the shape of an inverted T, whereby the inner leaf of blockwork and the outer leaf of brickwork are bedded side-by-side, up against the central, 70mm/2¾ in. wide vertical, foam-filled section of the Tee. Lintels used on non-cavity, interior walls of block- or brick-work, may also be of light-gauge steel, in the sectional shape of shallow, double-crank shaped channel – or they can be

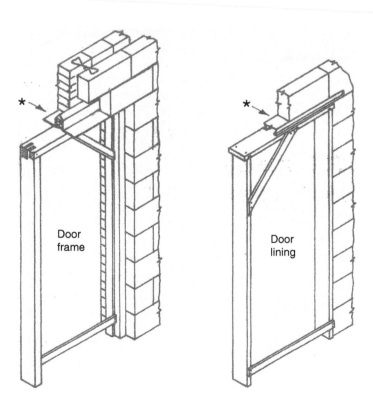

Door
frame

Door
lining

Figure 57 * Modern, galvanized-steel lintels above doorway openings

of pre-cast, reinforced concrete, called *plank lintels* – which are obtainable in various lengths. Note that factory-formed, pre-cast reinforced-concrete lintels are usually more reliable than concrete lintels cast in-situ (on site).

Modern methods of construction (MMC):   This term is a reference to prefabricated units (panels or component parts) of dwelling houses or apartment blocks, being made off site (in factories), then delivered to a prepared building site for erection – thereby speeding up the usual time-consuming site process. Although the term *Modern* methods of construction is being used, this technique started at the end of WWII, after 1945, when there was a desperate shortage of social housing. So, local authorities initiated the building of concrete-panelled and concrete-framed houses – as well as prefabricated, single-storey dwellings comprised of timber-framed, sheet-asbestos clad panels. The latter dwellings became publicly known as *prefabs* and exceeded their expected lifespan. But the concrete houses – although still around, in a much-reduced number – suffered *concrete cancer* (explained separately in this book) and other ailments related to *no-fines concrete* and *ferrous-metal corrosion*. Note that the exterior walls of a number of remaining concrete houses have now been covered with some form of cladding (no doubt to improve their bland appearance), but this puts another responsibility on the shoulders of surveyors and valuers – who need to identify a structure's core material and its likely lack of thermal-efficiency. And, making matters worse (from a detection point-of-view), it seems to be currently fashionable for DIYers to cover brick façades with horizontal shiplap/weatherboard cladding (of fibrous materials other than timber), in vogue with many fibrous, weatherboard-clad, upper-storeys of new-build houses.

Finally, modern-day methods-of-construction (re prefabricated, insulated units) are believed to be in the pipeline – although Swedish log-cabin homes are already established, in quality, longevity and triple-glazing. And, in recent decades, some interesting hybrid, two-storey or chalet-bungalows have been built. These comprise a prepared site with completed concrete-foundations and ground-slab, pre-designed factory-formed timber-studded panels, sheathed with plywood on their exterior faces – ready to be bolted down, bolted together and insulated to form the inner-leaf of the dwelling; first-floor truss-rafter roof assemblies (with the inner shape of Queen-post trusses, to accommodate the chalet's upper rooms) – and then an outer-leaf of face-brickwork is built around the structure, reinforced with right-angled stainless-steel frame cramps, screwed to the outer-face of the inner-leaf's plywood panels, leaving a 50mm/2 in. cavity. At a later stage, after the roof has been tiled and the dwelling is weathertight, the interior open-studded panels are filled with insulation, prior to dry-lining.

Moisture barrier: This is also referred to as a *vapour barrier*. Its function is to inhibit the diffusion of moisture-movement from one structural element to another. Materials used for this are usually of plastic- or polyethylene-sheeting.

Moisture content of timber: This refers to the necessary percentage of cellular-contained-water required to remain in timber components, in different environments, to enable them to remain in a healthy condition. For example, the initial moisture content (mc) of structural, first-fixing timbers, such as roofing timbers or floor joists, should not exceed 22% mc when first installed – and for second-fixing timbers, such as timber skirting-boards, architraves and doors, etc., the initial moisture content should not exceed 17% mc. And when newly-built buildings are dried out, these moisture-percentage figures should be expected to reduce to about 16% and 14% mc respectively. Note that, over time – especially if a building has been heated – these final figures should reduce by about 2 to 3%.

Moisture meters: Such hand-held, double-probe-ended, battery-operated instruments (as obtainable from such outlets as www.YorkSurvey.co.uk Supply Centre) are an essential piece of equipment for confirming suspected areas of damp-penetration in ceilings, walls and floors. Note that when checking interior wall- or ceiling-surfaces for dampness, care must be taken not to push in the pointed probes too deeply, thereby causing visual damage to the décor. At lower-wall level, the common surveying-technique is to rest the probe-points on the skirting's top edge, before gently pushing them in to the wall-surface. If checking a stained ceiling, one should again try to probe an area where staining or flaking-décor, etc, will obviate any probe-mark damage.

Monkey-tailed bolts: These are heavy-duty tower bolts, long in length and with protruding ball-peen-ended spiral shapes – which enable them to be operated vertically at the head and foot of tall doors, without overstretching or bending. They were traditionally used on large, industrial-type doors.

Monkey-tailed scroll: The name given to the shape of a handrail-scroll positioned over the bull-nosed or half-round curtail-step at the start/base of a geometrical staircase.

Mopstick handrail: A circular-shaped handrail of about 70mm/2¾ in. diameter, with a flat surface underneath of about 25mm/1 in. width, as a fixing surface for handrail brackets.

Mortar: This Latin-derived word simply refers to the common mixture of sand, cement and water with which bricks and building-blocks, etc., are laid and bonded. A similar mix might be used for plastering (rendering) walls; although (since the uptake of dry-lined walls), this is less frequently done nowadays. Also, *plastering mortar* is usually referred to technically as *rendering* – and as *coarse stuff* or *undercoat* in the trade. The other difference is that interior rendering and *setting* (the smooth finishing coat), consisted of an undercoat which was comprised of one part of putty-lime mixed with three parts of sand: (a 1:3 ratio).

   Bricklaying mortar is usually mixed to a ratio of 1:6 (one part of cement to six parts of soft sand); but traditionally, it comprised a 1:1:9 ratio of one part of cement, one-part lime and nine parts soft sand. *Pointing mortar (*for use on raked-out facing-brick joints) should be comprised of a 1:3 ratio of one part of cement to three parts of soft sand.

Mortar for hip-, ridge- and valley-tile bedding: See **Roofing-mortar**

Mortar-mix ratios: See **Mortar**

Mortise-and-tenon joints: These centuries-old framing joints are still widely used in the manufacture of woodworking joinery. Made by hand or machine, a typical use – as an example – is when the top rail of a panelled door is mortised and tenoned to the side-stiles. The thickness of the tenons formed at each end of the rail should be one-third of the rail's thickness (adjusted to the nearest machine-mortising, or hand-mortising chisel-size). But, the width of the tenons (a controversial subject in the woodworking industry), cannot be given their full width. This is because they need to be contained (held in the mortise) on their outer edges. So – as an example – master craftsmen are historically recorded to have reduced their top-rail tenon-widths by a half, but modern-day practice seems to favour only a one-third reduction. Although things inevitably change over time, one cannot ignore the evidence of historic panelled doors (framed up with primitive glues), still giving active service in many buildings, after hundreds of years of usage.

Mortise deadlocks: Such locks are without a quadrant-shaped striking latch (and are minus door-handles). They are only used, therefore, for security reasons – but are usually required (by Insurance Companies) to be fitted to traditional-type entrance-doors as extra security before they will consider issuing Buildings' insurance-cover on a dwelling. However, modern uPVC- or composite entrance doors are normally covered, as they are fitted with multi-locking side-shooting bolts, operated by the door-keyed lever handles. However, in the case of deadlocks, they are required to contain five levers. The more expensive deadlocks have a box-recessed striking plate/mortised *keep* – and the brass bolt contains two hardened steel rollers, to resist being cut with a hacksaw blade.

Mortise latches: These are mortised into the mid-area edges of internal doors (not requiring to be locked) and are either oblong or tubular-shaped. They require so-called latch lever-furniture, but are always supplied with a metal *striking plate* to receive the latch.

Mortise locks: These slim locks, which are housed/mortised into the mid-area edges of doors, are either **1)** traditionally 150mm/6 in. long and require a set of door-knob furniture and separate escutcheon plates to cover the keyholes; or **2)** modern mortise locks that are only 64mm/2½ in. long and require a set of *lock-lever door furniture*, with deep face-plates that provide their own non-escutcheoned keyholes.

Mould and bacteria growth in buildings: Technologically, every occupied dwelling is known to contain millions of mould spores – and in normal conditions, these remain dormant and the inhabitants are usually never aware of them. However, in favourable conditions of heat and humidity, mould spores grow and proliferate, producing unsightly and unhealthy-looking black patches of mould and bacteria growth to walls and ceiling-areas – usually between the wall- and ceiling-junctions initially. Over a period of time, this contamination will spread and is likely to contaminate clothes, bedding and furnishings, etc., as well as being a likely cause of diseases of the upper respiratory tract.

As explained in detail under the heading of **Condensation problems in dwellings**, written in this book, these problems can only be overcome if *ventilation, thermal insulation and heating* are fully understood and implemented.

Proprietary fungicidal-washes are sometimes used to tackle mould problems, but they are usually highly toxic and hazardous. Also, their effectiveness is only superficial and does not tackle the root cause of the mould problem – which is always in the fabric of the building, behind the decorated surfaces. Fungicidal washes kill the surface mould, but do not prevent it from returning. The treatment has to be repeated at frequent intervals ad infinitum.

However, a solution to mould problems is based on techniques developed by *MGC (Mould Growth Consultants) Ltd.*, many years ago, in the form of treatments using *RLT Halophen* (developed in MGC's laboratories and patented world-wide). The treatment, which involves three simple, straightforward stages, is claimed to be extremely effective, with a long history of success – and the product drying-time is almost immediate: **1)** All surface mould is first deactivated (by cleaning down with *RLT Bactdet 05*) to make certain that spores are not allowed to spread to other unaffected parts; **2)** Then *RLT Halophen* (a powerful, safe and long-lasting fungicidal membrane) is applied (coated) to the prepared surfaces of the walls and (if required) the ceilings. This acts as a barrier to prevent mould forcing its way through from the substrate. Both of the treatments (RLT Bactdet 05 and RLT Halophen), when applied, must extend 1m beyond all visible signs of surface-mould in each direction, to ensure capture of all spores and rooting systems; **3)** Then a *Biocheck* protective coating (in Matt or Silk emulsion paint) is applied as a protective coating – and to provide a decorative finish.

Note that all products by MGC Ltd are approved by the HSE under COPR 1986 (as amended) and are also approved for both amateur and professional use – with the proviso that the products must be applied in accordance with the manufacturer's instructions. Work can be carried out during occupation of the premises, if necessary – but it is recommended that all work is carried out in well-ventilated conditions and that vulnerable occupants, children and pets are kept away from the areas until all applications are dry.

Moulded bricks: **1)** This description either refers to the brickmaking process, or **2)** it may be used to describe special-shaped and moulded handmade bricks of the 15th–16th century, still to be seen on Tudor/Elizabethan buildings.

Mouse: There are a great number of sliding-sash windows still around, so the age-old *mouse* used for renewing broken sash cords is of great assistance in the re-cording operation. When a broken cord is being renewed, the sash has to be taken out of the boxframe window, and the *access pocket* has to be removed from the pulley stile to retrieve the broken sash cord with the balancing-weight tied to it. When renewing the sash cord, by feeding it through the pulley wheel at the top, the cord has to be weighted with a *mouse*. This aid is a hand-made device comprised of a string-line (a bricklayer's line is ideal) attached to a small piece of curved lead, about 75mm/3 in. long. Historically, an offcut of lead-sheathed electrical cable was used. It was slit open to remove the wiring and to receive and encase the end of the string line laid in it. With the replacement cord tied to the line's opposite end (via a clothes-hitch knot), the mouse is fed through the pulley wheel and lowered until it appears in the area of the opened access pocket. Then it is pulled through, removed and the end of the sash cord is tied to the top of the retrieved sash-weight. These are usually of cast iron or – on very old windows – cast lead. The weight can now be replaced through the pocket, ready for the new cord to be attached to the side-groove in the sash.

Muck: This historic bricklaying jargon, used at least in the London area, is a term meaning *mortar*. When working at scaffold heights and in need of mortar-replenishment, a bricklayer would shout out 'Muck up!' to the hod-carrying site-labourers below.

Mullions: See *Figure 3* on page 6: These are vertical, mid-area window-frame members (of timber or stone) positioned between the side jambs. They are normally twice rebated to receive fixed or opening sidelights.

Multi-ply: This plywood's high-rated strength is gained by virtue of its number of cross- plies being more than three – in multiples of two – starting with five plies. It is available in various thicknesses and superficial sizes. The reason for the odd number of plies in each and every board-thickness is to do with so-called *balanced construction*. If you think of an imaginary centre-line (also referred to in basic mechanics as the *neutral-axis line*) running through the sheet's thickness, the material on each side of that line must be the same to be structurally, equally-*balanced* with each other. Therefore, the centre-line neutral axis as related to the odd number of plies, enables the two *outer* plies to run in the same direction as each other – which is essential to avoid surface-distortion (warping) of the board. If, for example, after purchase, another layer of material – such as a *plastic laminate* – is bonded to one of its surfaces, it will require to be balanced on the other side as well.

Muntins: See *Figure 61* on page 199: These are vertical, mid-area panelled-door members positioned between – and side-grooved to receive – pairs of door panels.

N

Nails: Fixing devices have changed with modern technology in the last few decades and although nails, with some changes in design, are still driven in by a hammer, newly-designed T-headed nails are also being fired-in by strip or Coil Nailers (nail guns), using glue- or paper-collated strips of nails. But ***round-head wire nails*** are still widely used, obtainable in galvanized, sherardized and bright steel, with sizes from 25mm/1 in. to 200mm/8 in. The most commonly used sizes are believed to be 75mm/3 in. and 100mm/4 in; ***Lost-head wire nails*** are still available in galvanized, sherardized and bright steel. Their sizes range from 40mm/1½ in. to 75mm/3 in. The nail-sizes most commonly used are 50mm/2 in and 65mm/2½ in. for floor-laying with T&G boards. Note that with floor-decking, the traditional rule for nail-length is that they should be 2½ times the thickness of the decking material. Other nails still in use are: 3) *Brad-head oval nails*, 4) ***Lost-head oval nails***, 5) *Annular-ring shank nails*, 6) *Grooved-shank nails*, 7) *Masonry nails*, 8) *Panel pins*.

Nail-sickness: A term sometimes used to describe a slated roof displaying too many slipped-slates or *tingle*-repairs, thereby raising the question regarding the condition of the slate-nails, which may be excessively corroded and rusted away – especially if the wrong type of nails have been used.

National Building Specification: This title refers to a professional business in the UK – referred to as NBS – that provides written examples of detailed specification descriptions of work to be tendered for and carried out by contractors. These writeups are used by architects and designers, engineers and other professionals, to assist them in describing the work, the materials and the standards of workmanship to be used on their proposed construction projects.

Neat cement: Cement mixed with water-only, minus any sand or lime.

Needle: See *Figure 34* on page 78: A horizontal timber or steel beam, usually relatively short in length, that is inserted halfway through a pre-cut hole above a portion of brick wall to be removed (to form a doorway, etc.) and which is supported at each end by timber or metal props (such being technically referred to as *dead shores*). Once the opening has been formed, with a permanent beam or lintel inserted to carry the load above, the suspended brickwork is *made-good* and the props and needle(s) are eventually removed and the needle-holes are then made-good.

Newel caps: The tops of square-section newel posts are usually finished in three different ways: **1)** Shaped in themselves, such as **a)** with pronounced, chamfered edges all round, **b)** quadrant-shape rounded edges all round, **c)** with two cross-segmental shapes, **d)** with two cross-semi-circular shapes; **2)** With separate, recessed and side-projecting square caps, with machine-moulded edges – such caps being glued and/or nailed to the newel-tops; **3)** Separate spherical ornate caps turned on a lathe with projecting spigots, ready to be glued and inserted into predrilled holes in the top-ends of the newel posts. The three standard shapes that are commonly available in these shapes are referred to as **a)** *mushroom* cap; **b)** *ball* cap; and **c)** *acorn* cap.

Newel-drops: Usually, on any newel posts above ground-floor level, the newel post must project down, below its juncture with the acute-angled inner stair-string, to finish at a normally unspecified amount below the ceiling. If not specified, this drop should be not less than 50mm/2 in. and be given, at least, a chamfered-edge finish. Traditionally, these projections were known as *newel-drops*, or *pendants*, and – no doubt because of historic higher ceilings – they hung down deeper and were more ornate and often replicated the identical lathe-turned shapes of the tops of the newel posts above.

Newel posts: See *Figure 4* on page 7: Square-sectioned hardwood or softwood timbers, sometimes with lathe-turned ornamentation in their middle area, with a sectional size of not usually *less than* ex. 75mm x 75mm/3 in. x 3 in. and not usually *more than* ex. 100mm x 100mm/4 in. x 4 in. Their prime function is to structurally link the inner stair-strings (via the upper newels) to the landings and to support the handrails, whilst also providing a secure parallelogrammatic framework in which to fix the balusters.

NHBC (National House Builders Control) Warranty: See **Warranties on newly-built houses**

Niche: An indented shelf-recess in a wall, traditionally with a semi-circular arch above, to house a statue, or an ornament, etc.

Night latch: See **Cylinder lock (night latch)**

No-fines concrete: See **Concrete houses and bungalows**

Noggings: Short, horizontal timbers, usually of ex. 100mm x 50mm/4 in. x 2 in., that, when placed and fixed in a mid-area line between the vertical timbers of a stud parti-tion, stiffen up the individual studs and create a homogeneous unit.

Nominal sizes: This refers to the commercial, sectional sizes of sawn timber before planing.

Non-ferrous: This refers to metal used in building work that is non-corrosive (non-rusting).

Non-flammable: Something which is *not* easily set on fire.

Non-load-bearing walls: Although it may seem obvious if a wall, such as a garden wall, for example, has no load to carry other than its own weight, it is not so obvious with interior walls, when their tops are up against the ceiling. Frequently, such interior walls *are* non-load-bearing walls or partitions, but – even though they might only be of a half-brick thickness or of 100mm/4 in. blockwork construction – they could be carrying a load which is concealed above the ceiling – or concealed *within* the ceiling. So, take care in deciding (be it your responsibility) whether certain interior walls can or cannot be removed.

Norfolk latch: See **Thumb latch**

Nosing(s):   **1)** Small projections of stair-tread-boards, beyond the stair-riser-board's face, usually with a semi-circular, bull-nosed or round-edged splay finish; **2)** Similar finishes to the front- and return-ends of window boards; or **3)** similar small projections to the floor-line edges around the apron-lined faces of a trimmed stairwell.

Nosing line/Pitch line: See *Figure 69* on page 244: A theoretical reference to a stair's step-nosings, inferring that they should all conform by being up to the line (not short of it, or projecting beyond it) and must also adhere to this principle when a nosing-line is referred to as a *pitch line* (in stair-regulations) and turns (twists) radially around portions of tapered/winding steps.

Numbering and naming of floor levels in the UK and the USA: In dwellings and all other buildings in the UK, any floor that is wholly or partly below the surrounding ground-level – especially if part of that ground is up against the building – is usually classified as a basement. If it is wholly below the ground and wholly or partly sub-terranean (with ground up against the building), it may be classified as a cellar. The floor above a basement or a cellar, is the ground floor – and all floors above that are graduated as first-floor, second-floor, third-floor, etc. Contradictorily to the above, it is believed that the *first-floor* classification in the USA is given to the floor above a cellar or a basement, irrespective of it being at ground level. In other words, there is no reference at all to a *ground-floor* level. Note, however, that *ground floors* – regardless of their material construction – should be at least 150mm/6 in. *above* the outer-ground (or paving level), in relation to their DPCs/DPMs (damp-proof courses and damp-proof membranes).

O

O/A: A trade abbreviation for an *overall* measurement.

Obscured glass: Sheet glass moulded on one side to irregular or regular patterns, for the purpose of obscuring a clear view into a bathroom or toilet, etc. window. Of course, such glazed windows also obscure the outward view. Note that the Building Regulations'

Approved Document N1, which applies to dwellings, refers to *critical locations* where people are likely to come into contact with glazing and where accidents may occur, causing cutting or piercing injuries. These critical locations apply to (a) the glazing of doors and side panels between the finished floor level (ffl) and 1.5m above and (b) glass in internal and/or external walls and partitions between ffl and 0.8m (800mm) above. Therefore, any glazed windows, for example, that are less than 800mm above the interior ffl., will require to be glazed with safety glass – usually symbolized by a BS kitemark or an EN number.

Oil-based paint: Traditional paint made with a colouring pigment mixed with oil or varnish and thinners.

One-and-a-half-brick wall: This refers to a traditional, *single-leaf (skin)*, bonded wall of non-cavity construction, whereby the juxtaposition of the bonding-arrangement of the header-bricks on one side technically corresponds with the positioning of the stretcher bricks positioned on the other side to create either an English-bond or a Flemish-bond pattern. These 1½-brick walls – for obvious reasons – were sometimes referred to in the industry as *solid walls*. Pre-metrication, they were also referred to as 13½ inch walls; i.e., of 343mm thickness.

One-coat plaster: British Gypsum/*Saint Gobain* produce a one-coat plaster named *multi-finish* that can be applied as a rendering coat and finally trowelled to a finished surface – without the need to apply a separate finishing coat (as is/was the traditional practice).

One-pipe system: This is a reference to simplistic house-drainage, whereby just one main drainage pipe of 100mm/4 in. internal diameter is used below ground to collect and carry away soil water from WCs, waste water from sinks and wash-basins *and* surface water from rain. According to the drainage layout, the one-pipe/single-pipe system may involve many pipes jointed together, that may pass through one or more *inspection chambers* (formerly referred to as *manholes*).

Onerous covenants: See **Restrictive covenants**

Open-boarded fence: This description refers to a relatively modern fence, whereby it consists of the usual fence-posts, but the three horizontal rails are of a rectangular section of at least 75mm/3 in. x 32mm/1¼ in. treated timbers. These are mortised into the posts, and the treated boards (say, of a 100mm/4 in. x 18mm/¾ in. section) are nailed vertically to the rails, in a hit-and-miss staggered pattern, on each side of the fence. The overlapping between the edges of the boards (as they relate to each other on each side) is discretional, but seems to be about 25mm/1 in., dependent upon the degree of neighbourly privacy required.

Open-ended eaves: See *Figure 29* on page 68: This refers to a historic form of eaves' projection – still to be seen on 18th-/19th-century properties – which has neither fascia boards nor soffit boards. The open-ended rafter-projections are usually bullnose-ended or half lambs-tongue shaped; and the partially glimpsed underside of the roof is over-laid with thin, bead-edged sarking boards – which do not usually clad the whole roof. As there is no fascia board upon which to fix the guttering, it is supported by ladle-shaped wrought-iron bars, screwed to the sides of the projecting rafter-ends.

Open-grained timber: **1)** This can be a reference to timber which has been affected by alternating wet-and-dry weather conditions over time; or, **2)** It may also be used to describe certain species of timber that naturally have a coarse-textured grain.

Open stair-string: See *Figure 32* on page 75: This refers to a traditional staircase which has its step-shapes cut away on the top edge of its outer-string to support the ends of its protruding, nosed-treads and the ends of its rebated-and-mitred riser boards.

Open stairwell: This technical description usually describes successive flights of stairs with quarter-turn or half-turn stairs, or landings within a stairwell that serves a number of storeys – and thereby creates a clear, *open stairwell*.

Open-valley (with lead-lined gutter): See *Figure 76* on page 262:   This type of roof-valley junction requires two sawn boards with a sectional size of 225mm/9 in. width and a thickness of 25mm/1 in. to be laid and fixed adjacent to each other in the vee-shaped valley recess, after their adjoining edges have been bevelled (planed) to meet the vee-abutment. Once they are fixed to the *cripple rafters* on each side of the valley rafter, tri-angular timber *tilting fillets* (of ex 75mm/3 in. x 25mm/1 in. sawn) are then fixed to each top edge. The purpose of the fillets is to create a small vertical upstand on each side of the inverted valley-boards, upon which the angled valley-tile edges will sit and protrude above and into the open-valley gutter. The tilting fillets also allow the lead-lined gutter to be *dressed* up their vertical faces and dressed over their top surfaces – thereby pro-viding a safely-weathered seating for the roofing-underlay and the tiles.

   Once the open valley has been lined with a heavy-grade lead – laid with overlaps of at least 75mm/3 in. across its width – a breathable, impervious membrane (such as Tyvec Supro or Marley Supro) must be draped over the rafters, beneath the tiling battens, and neatly angled-cut to protrude slightly over the edges of the lead-covered tilting fillets.

   One-and-a-half-width and common-width plain tiles are then hung on the tile battens at the valley's edges, to be marked for cutting to the angle of the tilting fillets' edges and with consideration for creating a necessary arrangement of staggered side joints.

Orangery: See **Conservatory/Orangery**

Oriel windows: These traditional, small windows protrude from the face of buildings, but do not take any support from the ground; they are usually supported by corbels or brackets.

Oriented Strand Board (OSB):   This resin-bonded sheet-material (comprised of densely compressed sliced-wood shavings), has a popular sheet-size of 2.4m x 1.2m/8ft x 4ft. and popular thicknesses of 12mm/½ in. and of 18mm/¾ in. As sheet material, it can be used for cladding, etc. (replacing plywood on certain jobs). It is also produced as T&G (tongue and grooved) flooring panels, acting as subfloors.

Ordnance Bench Mark (OBM):   In all site-levelling operations, it is essential that the various levels required should have reference to a fixed datum. In a rarely available, ideal situation, this datum would be the Ordnance Bench Mark (OBM) – which is an inverted V-shaped, chiselled-out arrow-head with a horizontal recessed-line above it. These are carved in stone blocks which are usually found set in the walls of public buildings and churches. The inverted horizontal centre line at the point of the arrow is a national land-level reference point related to the *mean* (standard) sea-level measured at Newland in Cornwall, UK. And this was originally calculated by the ordnance surveyors.

Note that OBM readings are recorded on Ordnance Survey maps related to block plans of built-up areas and can usually be viewed or bought from a local authority's Building Control Department. Where there are no OBMs in the vicinity of a proposed building site, an alternative and reliable datum must be used. These might include the stone step of a building, the top edge of a nearby building's stone plinth or the top of a cast-iron manhole/inspection cover in the road.

Ordnance Datum: See **Ordnance Bench Mark (OBM)**

OSB: See **Oriented Strand Board (OSB)**

Outer skin: A trade reference to a wall built on the exterior-side of a cavity wall, which may be of brickwork, or (for cost-saving reasons or aesthetic preference) of sand-and-cement rendered and painted blockwork. Nowadays, however, still with cost or preference in mind, the blockwork may be of ready-mixed silicone-coloured, polymer-modified render or modern cement-board shiplap, coloured cladding.

Outer string: The long board-on-edge that houses all of the step-ends of a staircase at the open side (away from the wall) of a typical flight of stairs.

Over-boarding to ceilings: This is a reference to relining (covering over) the underside of existing ceilings, usually with abutting and staggered sheets of plasterboard. After locating the underside-edges of the concealed ceiling- or floor-joists and screwing up the boards, the joints should be scrimmed and caulked – or scrimmed and skimmed (set) with plaster – prior to redecoration. Over-boarding ceilings is a remedial action used commonly nowadays as an alternative to removing defective lath-and-plaster ceilings. Anecdotally, it is also alleged to be done to cover up Artex-coated ceilings suspected of containing ACM (asbestos-containing materials). This is a practice which is not supported by Building Regulations' Control, unless a notice is displayed nearby to inform future occupiers of the premises, of such concealment.

Oversite-concrete: In its simplest form, this term refers to a layer of concrete of at least 100mm/4 in. thickness laid over the levelled ground of a new dwelling to seal the earth.

P

Pad-and-beam foundations: See *Figure 47* on page 118: Foundations for all forms of structures are extremely important and their final design is the domain of a structural engineer. But, different types of foundation might be considered and, for certain relatively lightweight, single-storey structures, pad-and-beam foundations have a certain appeal. As their name implies, they are comprised of cube-shaped pads of concrete measuring 686mm/2 ft. 3in. in length and width x 230mm/9in. depth. Brickwork piers of 225mm/9 in. square are built on these and similar-sized square-shaped, reinforced concrete beams are cast in situ on their top, load-bearing surfaces – ready to support whatever floor may be designed within the regulations.

Paddle-shaped tread stairs: See **Alternating-tread stair**

Padstones: These are pre-cut, natural blocks of stone or pre-cast concrete blocks which are placed centrally under the end-bearing points of heavy beams (such as steel girders/RSJs), to help spread the point-load on load-bearing walls.

Painter's eighth: This traditional term refers to the painting of glazed, wooden windows, regarding the bevelled putty-beads, whereby, to guard against the drying-shrinkage of the putty and to create a watertight seal, the finished edges of the paint would be carefully lapped onto the glass by approximately *one-eighth* of an inch, (⅛")/ 3mm. This practice was referred to as '*adding a painter's eighth*'.

Paint remover: A solvent liquid which softens oil-based paint or varnish, making it easier to remove by scraping and/or brushing.

Pallets: Traditionally, pallets (small pieces of softwood measuring about 7mm/¼ in. thick, x 50mm/2 in. wide, x 100mm/4 in. long) were built into the bed-joints of brick-reveals by bricklayers to provide fixings for the eventual door-linings, frames or panelling. Nowadays, such fixings are made directly into the masonry substrate with so-called *frame-fixing screws*, after having drilled slightly-lesser diameter pilot holes into the brick- or blockwork.

Panel adhesives: For at least two decades now, panel adhesives (their generic reference, but commonly referred to as *grab adhesives*), have taken over from mechanical fixings with nails and/or screws, etc., of a number of second-fixing items such as skirting boards, window boards, wide-architraves (for fixing their outer, underside edges), etc. Some of these grab adhesives are solvent-free and gap-filling. Others are suitable for exterior use and some are suitable for both interior and exterior use. The adhesives are usually in 350ml cartridge tubes, which fit into most cage-type sealant guns.

Panel-bolt connectors (as used for kitchen worktop-joints): These simplistic bolts (two or three of them) are used for the non-simplistic task of joining the routered *end* of one kitchen-worktop to the routered *side* of another. Stub-ended, routered T-shapes are made in the two, underside joint-edge surfaces of the worktops – ready for gluing and bolting up. Numerous, corresponding slots are made in the joint-edges (with a portable biscuit-jointer), in readiness for gluing up. The elliptical-shaped, feathered-beech biscuits are inserted, the special glue is applied, the bolts are inserted and speedily tightened up.

Panelled ceilings: *Figure 58*: There are two analytical questions that need to be raised in one's mind regarding existing panelled ceilings: **1)** Are they original to the property? **2)** If not, why have they been panelled? Historically, plaster-moulded panelling (into squares or oblong shapes) on lath-and-plaster ceilings with elaborately corniced

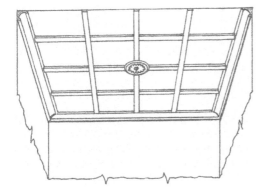

Figure 58 Pictorial impression of a timber-filleted, panelled ceiling, fitted against cornicing and a central ceiling-rose

perimeters, were quite commonly used in public buildings and large houses – but not usually in common dwelling-houses.

So, when appraising a dwelling-house ceiling that has been multi-panelled with door-stop sized timber fillets – usually with a sectional-size of ex. 50mm/2 in. x 12.5mm/½ in., or with astragal-moulded fillets (scribed to each other at their multi-shaped junctions) – the first thing to do is to sound-tap the ceiling with the soft, balled-up karate-side of your hand, to find out whether it is of **a)** lath-and-plaster, **b)** plasterboard, or **c)** a more lighter-weight building board, such as fibreboard, *asbestos sheet material* or hardboard. Such soundings, with a little experience, should easily distinguish one from the other. Then you need to question your findings in the following way:

a)   Cracked lath-and-plaster ceilings are quite common, such being a sign of the aged bovine-haired mortar breaking away from the laths and threatening partial-collapse. So, to offset this threat and the expense of messy ceiling-replacement, house-owners sometimes have the ceiling panelled, as a means of holding it up. Alternatively (or additionally), cracked lath-and-plaster ceilings are also lined with heavily-embossed anaglypta paper, primarily as a means of hiding the cracks, but incidentally having a proven benefit of holding up the ceiling.

b)   Non-plastered plasterboard-and-panelled ceilings might indicate that a defective lath-and-plaster ceiling has been over-boarded with plasterboard, but not skimmed with a coat of finishing-plaster, thereby necessitating their abutment joints to be hidden within a pre-designed pattern of panelling members.

c)   This is no doubt the most common panelled ceilings to be encountered on Building Surveys of dwelling houses – and initial sound-tapping usually discovers them easily. Such ceilings are usually the result of extensive partial-collapse, leading to complete removal and temporary repair with lightweight materials. Some of these 'temporary' repairs, made as a result of war damage in the WWII years, are still waiting to receive 'proper' ceilings – and others amongst them might have been replaced on the cheap by their owner-occupiers.

The main concern here, is that if they used thin ACM (asbestos-containing material) sheets, they should not be disturbed and must be reported upon or checked out by a specialist.

Panelled walls: Since many hardwoods (the deciduous species of timber), like oak, are now very much depleted in most countries, panelled walls (as used in the Houses of Parliament and other stately buildings and grand houses) are not seen in building design nowadays – which might also be due, of course, to cost and modernity. However, wall-panelling – similar to hardwood panelled doors – was pre-formed in joinery workshops, but transported to the site in pieces, ready for site-assembly and fixing to the unplastered, timber-grounded walls and doorway-reveals. Such panelling was either built to dado-rail-height, at 900mm/3 ft. above floor-level, to the doorway-height of the architraves, or to the ceiling-height, finishing with frieze rails and hardwood built-up cornices at ceiling level.

Panic bolts: Such bolts are commonly used on double doors serving fire exits to theatres and supermarkets, etc. *Downwards* pressure on a mid-height, mid-jointed horizontal locking-bar, releases the outwards-opening doors. Traditionally, panic bolts operated by applying an *upwards* pressure on the locking bar – but this was (sadly) discovered historically not to operate as well in a panic situation – as a downwards pressure.

Pantiles: Such roofing tiles have two S-shaped corrugations (one concaved, the other convex) that overlap and interlock with each other at their sides and form gutter-like channels up the roof, at right-angles to the eaves.

PAR: This is a traditional reference to timber being ordered as ***planed all round***, i.e., planed on each of its four sides.

Parapet: See *Figure 14* on page 30: The uppermost part of a building's outer wall that rises to a low height above the eaves' edge of a roof to form a *parapet wall* – against which is built an unseen *box gutter*, also referred to as a *parapet gutter*. Note that parapet walls (usually of 225mm/9 in. single-skin construction) are exposed to extreme weather conditions on their top, face- and back-sides – and are therefore prone to descending, penetrating damp.

Parapet gutter: See **Box gutters**

Parging (or pargeting): 1) This was the term used for the rough rendering/plastering that was historically applied as a lining by bricklayers to the internal brick-surfaces of chimney-flues as they were being built. The bricklayer's lime-mortar was used for this and it was applied with the bricklaying trowel in incremental stages as the chimney stack was being built – such stages being governed by the workable-reach of the bricklayer's arm and the awkwardness of a mortar-laden, upside-down, hand-held trowel. Hence the term *rough* rendering/plastering used above – which inevitably aided the build-up of soot in traditional chimney flues, especially on the reflex- and obtuse-angle-shaped bends, which aided the deliquescent soot-ash problem explained under **Deliquescent soot-ash staining to chimney-breast plaster; 2)** The term parging or pargeting is/was also used to describe ornamental, exterior plasterwork.

Parliament hinges: These were/are also referred to as *H-hinges* or *shutter hinges*. The reference to the letter H gives a line-diagram image of a parliament hinge opened up, by imagining a vertical knuckle in mid-area of the H's cross-bar. The two uprights of the H are the opened-up, screw-holed fixing leaves of the hinge. The purpose of the hinges is to allow an open door – or shutter – (if required) to lie as flat as possible against a wall, or against projecting architraves and skirting boards, thereby requiring the hinges' knuckles to be extended more than is usually required. Knuckle projections are usually obtainable in incremental sizes.

Parquet flooring: 1) A finished floor-covering of hardwood or softwood (redwood) blocks or strips, laid to a geometrical, ***herringbone*** pattern, and traditionally bonded to a sand-and-cement screeded floor. Such blocks were/are usually of 19mm/¾ in. thickness; **2)** Nowadays, parquet flooring is also available as an overlay floor, which uses much thinner strips of timber or a hard-wearing type of vinyl-lay.

Particle board: This is a general reference to manufactured sheet material such as chipboard, MDF (medium density fibreboard) and OSB (oriented strand board), etc.

Parting beads: These are thin strips of round-edged wood, fixed into shallow grooves on the faces of the opposing *pulley stiles* of a traditional boxframe-window. Essentially, parting beads create separate channels to keep the side-by-side upper and lower sliding-sashes apart during the opening-and-closing operation.

Parting slip: See **Wagtail**

Partitions: See **Metal-stud partitions** or **Timber-stud partitions**

Party wall: An interior, dividing wall (with a nominal 225mm/9 in. thickness) between semi-detached buildings and terraced houses. Each property owns and is responsible for a theoretical half-thickness of such walls.

Party Wall Act 1996:   This applies in situations where work is to be, or is being, carried out to a party wall that is either part of a building, or part of a boundary wall. **1)** It provides a pathway for negotiations to take place regarding the construction work to be carried out; **2)** It covers any likely situations regarding a need to excavate below the foundation level of any adjacent building, or a structure within 6m/19ft 8 in. of the boundary; **3)** It provides for owners to carry out work on a party wall, if they have served the necessary notices.

There is a mechanism in the Act which allows for the appointment of a surveyor to act as an adjudicator, if any disputes arise between the parties involved.

Passive-stack ventilation: This refers to circular, insulated air-ducts that rise up through a dwelling from a ceiling terminal at ground-floor level, enter into the roof void (either vertically or via maximum 45^0 bends) and exit through the roof to the ridge height. Mastic seals must be provided around the ducts at the points of entry and exit at floor- and ceiling-junctions. Easy-to-clean ceiling terminals must be provided with open areas equal, at least, to duct-diameter and with automatic or manual control.

Patina:　A protective film of oxide which forms naturally on metals exposed to air –
such as the green patina on copper.

Pattern staining:　See **Plasterboard discolouration**

Pavement lights, prisms or **windows:**　Historically, such *windows* – usually up against,
or close to front, elevational walls – were comprised of small, solid glass blocks (prisms)
set closely together in multi crisscross-barred, cast-iron frames. They were used above
basement storage-rooms of shops, to let in daylight. Many pavement lights are still in
existence – often seen to be in need of repair/reglazing.

Pebble-dashing:　This traditional plastering technique was used on the faces of exterior
walls of houses for two reasons: **1)** Because the outer walls had been built with *common
brickwork* or *blockwork* for cheapness, with the intention of a pebble-dash *veneer* to
provide the finished appearance; or, **2)** because the aged face-brickwork needed some
attention, such as raking out and repointing; stitching (a technique for repairing cracks);
the cutting out of eroded or *spalled* bricks and replacing them, etc. – all or any of which
would no doubt have been much more expensive than pebble dashing.

Pebble dashing is comprised of two coats of sand and cement rendering to a 3:1 ratio
mixture and an approx. thickness of 9mm/⅜ in. for each coat. The first coat should have
a waterproofer added and – after being keyed with a *wire scratcher* – it should be left
for about three days. The second, *dashing*-coat, should have a plasticizer added to it (to
make it fatty). Pebble types vary, as do their sizes from 6mm/¼ in. to 25mm/1 in.

Pebble-walled houses:　Historically, the outer walls of such houses were built with large,
rounded pebbles bedded in mortar – but in the 19th century, the building-technique
changed; seemingly influenced by cast-in-situ concrete and stone. To create the mould
box for each wall, brick- or stone-quoins (outer corners) and doorway-reveals of a
building were built and (after setting), timber-shuttering boards were incrementally fixed
to their front- and rear-faces. Pebbles were built-up in a dry state against the shuttering
on each side – and then the remaining middle-area between the two pebble façades was
filled with concrete. The tamping of the concrete squeezed in between the pebbles to
create a good bond. This process continued in incremental rises of the shuttering and
the wall, until the full storey-height was reached. After the walls were built, the joints
between the pebbles were pointed with a sand-cement-and-lime mortar.

Pediment:　A European-influenced, Classical ornamentation featured above doorways
and windows, etc., usually in a closed- or open-topped, 30⁰ triangular, moulded shape.
Still in good evidence in cities and towns in parts of the UK, either formed in hardwood
on interior design, or stone on exterior work. A few notable architects introduced this
ornamentation into the capital cities in the 16th/17th centuries.

Peg tiles: These early-19th-century, plain tiles, handmade with clay, are pierced with two fixing holes near their tops, enabling them to be fixed to the roof with approx. 6.5mm/¼ in. diameter oak pegs (or roofing nails).

Pellets: Small, cork-shaped wooden plugs (where the grain-direction runs *across* the cork-head), that are used for patching/pelleting counter-bored screw-holes used in screwing and pelleting quality *door-linings* (as an example).

Pelmet: A thin, traditional fascia board – or a width of cloth material – placed above the head of a window, sometimes with ornate features, to conceal the curtain track and fittings.

Pendants: a) A reference to *hanging light-fittings*; **b)** See also under **Newel-drops**.

Penthouse: 1) An apartment built on the *flat-roof* of a block of flats, traditionally with walking or garden space around it; **2)** Nowadays, this term is used to refer to an upper-most apartment (*on the top floor*) of a block of flats.

Penthouse roof or **Pent roof:**   Such a roof usually slopes in one direction only, similar to a so-called *flat roof* (which usually has a pitch of 5 to 10^0), but with a pitch similar to a shallow lean-to roof, with a pitch of about 15^0.

Pergola: A sturdy, open-framed, two-sided and beam-topped structure in a garden, used as a form of trellis for climbing plants.

Permitted Development: 1) When building an extension, planning permission is not required, as long as the ground area covered by such an extension and any other buildings within the boundary of the property – excluding the original house – is not more than half the total area of the property; **2)** Any part of the extension is not higher than the highest part of the roof of the existing house; **3)** The eaves of the extension are not higher than the eaves of the existing house; **4)** Any part of the extension does not extend beyond any wall facing a road (i.e., the so-called *building line*), if it forms the principal- or side-elevation of the original house; **5)** The eaves are no more than 3m/9 ft.10⅛ in. in height if any part of the extension is within 2m/6 ft.6¾ in. of the property's boundary; **6)** The materials used in the exterior work – except in the case of a conservatory – are of similar appearance to the existing house; **7)** Any upper floor window on a side elevation within 15m/48ft.2⅜ in. of a boundary with another house is obscure-glazed – and is non-opening, unless the parts which can be opened are more than 1.7m/5ft.7in. above the floor of the room in which the window is installed; **8)** A side extension does not exceed 4m/13ft.1½ in. in height, or be wider than half the width of the original house; **9)** In a single storey extension: **a)** the extension does not extend beyond the rear wall of the original house by more than 4m/13ft.1½ in. for a detached house, or 3m/9ft.10⅛ in. for any other type of house; **b)** the height of the extension does not exceed 4m/13ft. 1½ in.; **c)** no part of the extension is within 3.5m/11ft.5¾ in. of any property boundary with a road opposite the rear wall of the house.

Note: Although the permitted development described above from a gov.uk planning portal, lists criteria to be met to not require **planning permission** before carrying out any such work, you will still need *Building Regulations' approval* from your local Council's **Building Control Department**. See **Building Regulations Control and Approval** described herein under that heading.

Perp- or **perpend-joints:**   This is bricklaying-trade terminology for perpendicular mortar-joints between bricks and/or building-blocks, etc.

Personal Protective Equipment (PPE):   Various items (such as safety helmets, goggles and footwear, etc.) are required to be worn on building sites nowadays.

Piano hinge: See **Strip hinge**

Picket fence: A low-rise fence comprised of posts, top and near-bottom rails and closely-spaced *pickets/stakes* with spear-pointed, segmental, or semi-circular-rounded tops. The pickets are usually of narrow battens with a 63mm/2½ in. x 20mm/¾ in. section.

Picture rails: Traditional, ear-shaped wooden rails, fitted and fixed around the walls of a room, usually at the level of the top of the architrave-head of a room's doorway. Whatever the height, the picture rails should be kept parallel to the ceiling, regardless of actual levels.

Pier(s): 1) In conversion work, when the ends of a load-bearing beam or a steel joist (RSJ) are to be seated on half-brick-thickness walls or 100mm/4 in.-thickness block-work partitions at each end, projecting brick-piers – bonded into the wall – are needed to give the walls extra thickness and stability at the points of increased load; **2)** Such piers are also used to strengthen and stabilize the corner-quoins and mid-length points of half-brick-thickness walls of a certain length – as can be seen ubiquitously in brick-built garages.

Pig-in-the-wall: This rather odd, historic bricklaying-phrase refers to shoddy work in bricklaying, whereby a length of wall being built has reached a certain height at each end (before the wall has been built up in between), when it is discovered that the number of brick-rises laid at one end (the corner quoin) is more than the number of brick-rises laid at the other end! Such an anomaly is referred to as having a *pig in the wall* and it can happen when bricklayers do not use a *gauge rod* (a storey-height timber batten, marked with equal brick-rises on it) – which should be used regularly from a fixed datum to check the incremental rise of the brick quoins at each end.

This is more likely to happen when a wall is being built by two bricklayers, one at each end. One making the mortar-bed joints slightly too thin, the other making them slightly too fat. Over a certain number of brick-rises, this soon creates a problem – to such an extent, that the extreme end-heights might even be level to each other and yet one end will have more brick-rises than the other!

One of the reasons why this anomaly goes unnoticed for a while, during the building process, is because the practice of building up each quoin first is being employed. This leaves the middle area to be filled in at a second stage. So, the quoins are built with plumbness to their end- and face-sides – and are *raked back* (stepped up) towards the quoin by a half-brick length reduction on each course. It might only be when the middle area is being filled in, that the pig-in-the-wall will be discovered. If the wall being built is of common bricks, to be plastered or covered up during the finishing stages of a contract, the anomaly might be levelled up with a split (tapered) course, but, in the case of face-brickwork, this is usually not acceptable, so dismantling and rebuilding would have to be carried out.

Pilasters: Traditional, rectangular, shallow piers, with fluted or reeded facework, plinths at their base and capped at the top, that project from the face of a wall – usually on each side of a doorway. Such ornate embellishments were formed in stone, timber and plasterwork.

Piling: This is a form of underpinning to support reinforced-concrete beams (acting as foundations for a building) that are subsequently cast upon their tops. Varying diameters and lengths of bullet-headed steel-tubes are mechanically driven into the ground and subsequently filled with a steel-reinforcing cage and a prescribed mix of concrete. As the concrete is being poured, the steel-tube former is gradually withdrawn mechanically – and the weight of the wet concrete settles into any small voids in the ground, which adds to the pile's load-bearing strength, via the ragged side protrusions.

Pillar taps: These traditional, head-turning taps have a tubular threaded-valve that fits through a hole in a wash-hand basin or sink – and if the sink is of thin, stainless steel, there may not be quite enough thread for the back-nut to be tightened, so a fitting known as a *top hat* (with a hole in its top) has to precede the back-nut to ensure a tight connection.

Pinch rod: This term refers to a traditional technique for checking the precise height or width between floor and ceiling, or between one wall and another, etc., or between any enclosure, by using two overlapping rods (timber battens), *pinched* together by hand in their mid-area and extended till their ends touch the opposite surfaces. Then one or two pencil lines are marked across the double-rodded mid-area to enable the rods to be reassembled when laid on the material to be cut.

Pin hinges: These are butt hinges that have dome-head pins inserted through their crenellated knuckles, to enable a door (after being hung) to be removed without removing any screws. To facilitate this, the knuckles need to project slightly more from their usual housed positions.

Pitch: The angle of inclination to the horizontal plane of a pitched roof or a staircase, etc. For example, dwelling-house stairs should not have a pitch of more than 42^0.

Pitched-roof angles: To suit the weathering requirements of most available tiles and slates, etc., isosceles-shaped pitched-roofs do not usually have pitch-angles of less than 25^0.

Pitch-fibre drain pipes: These 100mm/4 in. diameter pipes, made from coal-tar pitch and asbestos fibres, etc., were used to some extent for surface- and foul-water drainage in the mid-20th century, until it became known that they were not ideal for their purpose. They were found to deform if subjected to ground-pressures and also suffered damage from rats gnawing through them. If such pipes are detected on a Building Survey (usually their black outlets can be glimpsed where they join the open, half-round earthenware channel in the inspection chamber), their existence should be reported in the survey report, with a recommendation for a CCTV survey to be carried out by a Drain Specialist.

Pitch line: See *Figure 69* on page 244: This is described in the Building Regulations as a notional line used for reference to the various regulatory rules; it connects the nosing-edges of all the treads in a flight and also serves as a line of reference for measuring the 2R+G rule on tapered-tread steps.

Plain tiles: See **Roof tiles and slates**

Plank lintel(s): This term is a reference to relatively modern, pre-cast, steel-reinforced concrete lintels that presumably gained their name by resembling a thick timber-plank. However, this name helps to distinguish them from traditional, deeper concrete lintels.

Planning permission: Apart from building maintenance, repairs and redecoration, the building work that requires liaison and authorization with a local government authority, comes under three categories: **1)** *Planning permission*; **2)** *Building Regulations approval*; and **3)** *Permitted Development*.

Categories **2)** and **3)** are covered herein, but it must be understood that there is a degree of interaction between the three categories. Planning-permission control by each local government authority is made up of a body of councillors (guided by their technical officers/surveyors), who use preconceived planning-strategies in relation to observance of individual building-designs and structures, and in relation to infrastructure, amenities and future plans. The planning authority also has to take conservation areas and environment issues into account. However, planning permission (or outline planning permission) has to be supported by Building Regulations' approval of the structural-design details and the detailed specification of the intended work.

Planted moulds or **moulding:**   See *Figure 12* on page 26: This woodworking-term (*planted*) refers to a separate moulding fixed to another component by nailing or panel-pinning, as opposed to *stuck* mouldings that (confusingly) have been formed onto timber edges by moulding machines – not by being adhered with an adhesive!

Plaster angle-beads: These relatively modern plastic or galvanized angle-beads are used to form 270^0 quoin-angles on internal and/or external walls. Different sectional sizes are needed for 3mm/⅛ in. finishing plaster on plasterboard, or for thicker rendering coats.

Plasterboard discolouration: This is also referred to as ***pattern-staining*** and can present itself on the surface of skimmed- and/or unskimmed-plasterboard ceilings and studded walls – although it seems to occur mostly on ceilings – in the form of a regular pattern of narrow, dark-shaded/stained areas, depicting the width and spacings of the unseen undersides of the timber joists above. This could also happen if the plasterboard is in contact with the underside of a steel beam, and the discoloration is usually caused by the temperature differences on both sides of the ceiling or studded wall.

Plasterboard screws: Instead of galvanized/sherardized nail fixings, nowadays plasterboards are fixed to the undersides of ceiling joists – or to the face of stud partitions – with bugle-headed, phosphate-coated screws, quickly driven in with auto-feed screwdrivers. Note that the bugle-shaped heads tend not to damage/tear the paper surface of the boards.

Plasterboard sizes: Standard sheets of plasterboard, most commonly used for ceilings, stud-partitions and dry-lined walls, are sized 2.4m x 1.2m x ***9.5mm***. These are to fit joist- or stud-spacings of 400mm centres; or the boards are of ***12.5mm*** thickness for joist- or stud-spacings of 600mm centres. It must be understood that these metricated lengths only suit present-day ceiling-joist spacings of 400 or 600mm centres – they do not suit pre-metric, imperial joist-spacings of 16 in. (406.4mm) centres. There is a choice of tapered- or square-edge boards. The former being required for dry-linings (to allow for a wide band of filling), the latter for a skimmed/setting coat of finishing plaster, after the 3mm/⅛ in. open joints have been reinforced with self-adhesive fibreglass-mesh tape.

Therefore, if replacement ceilings are needed for pre-1970s properties (built prior to metrication), 2.5m-length sheets of plasterboard would have to be used and cut to suit the imperial-sized joist-spacing centres of 16 inches x 6 spacings = 96 in. (i.e., 2. 438m/ say 2.4m). This only creates a minimal waste of 62mm per sheet (to centralize the plasterboard-joints on the joists or studs) – but another anomaly is that the 2.5m-length sheets are only produced in 12.5mm thicknesses (not 9.5mm).

So, this could affect *small-works' repairs*. Metric-equivalent plasterboard sizes were produced for a considerable number of years, but the British Gypsum company (or their French partners, Saint-Gobain) have now stopped producing imperial-equivalent sized sheets of plasterboard sized 2. 440m x 1. 220m, no doubt because of a relatively low demand.

Plastered (interior) walls: Since *dry-lining* techniques on brick- and block-built walls have evolved over the last four or five decades, so-called *two-* or *three-coat work* with built-up layers of wet plaster-mixes has mostly disappeared. Although, a certain amount of such work is still required to be done on maintenance and repair work, which is usually reckoned to outweigh the amount of new works carried out in the UK. However, it has to be realized that wet-plastering techniques, producing *solid-plastered walls*, are extremely labour-intensive and therefore costly.

Traditionally, then, solid-plastered two-coat work consisted of a *floated* (*rendered*) coat of rough-textured plaster, with an approx. thickness of 13mm/½ in., which was used for truing up and keying the surface. This was followed (soon after *setting* had occurred) by a 3mm/⅛ in. approx. thickness of skimmed *finishing plaster*. However, there is also a modern-day change to this procedure – relatively small solid-plastering jobs can now be done with one-coat multi-finish plaster, which – after being trued up as a floated/rendered coat, can be further trowelled – after its initial setting action to bring it to a smooth, finished surface.

Plastic wood: Patented paste made of cellulose, wood dust, resins and plasticizers, and other ingredients and solvents. It can be useful for repairing wood prior to repainting, or for filling small holes in wood.

Plasticizer: See **Additive or Admixture**

Plate(s): 1) Horizontal timbers that hold the ends or edges of vertical, inclined or level timbers in a state of alignment and framed-spacing, as in *stud-partition floor- and head-plates, pitched-roof wall-plates* and *floor-joist wall-plates*, etc. **2)** Metal components such as *striking plates* for door-locks and latches.

Plinth: This building term usually refers to a flat-faced projecting band at the base of exterior walls, formed with a rendering-mix of not more than a ratio of 3:1 (3 of semi-coarse sand to 1 of cement). Plinths are usually thought of as visual features (and may have a moulded top edge similar to skirting boards, or a splayed, weathering-edge), but – like skirting boards – they are mainly there to protect the base of the wall from damage; however, unlike skirting boards, they also protect the wall from rain-splashing. The sectional size of plinths can be seen to be from 150mm/6 in. to 225mm/9 in. and above. Note that (a) if a wall has a DPC (damp-proof course), the top edge of the plinth should not be above its bottom edge, and (b) if a DPC cannot be seen or detected, it is possible (in old-build properties) that a DPC does not exist.

Plinth blocks: See **Architrave/Plinth blocks**

Plot: An area of land designated for building development.

Plugs: See **Wall-plugs**

Plumb-cut: This is usually a reference to an acute-angled, vertical sawcut made at the head of roofing members, such as rafters and purlin struts, etc.

Plywood: This *sheet material* is obtainable in various thicknesses from 6mm/¼ in. (which is believed to be one of the popular thicknesses) and is referred to as *3-ply* – other thicknesses are available in *5-plies*, *7-plies* and *9-plies* and of 18mm/¾ in. thickness; these three are referred to as *multi-ply* boards (or sheets).

Technically, plywood is referred to as having a *balanced construction* of odd-numbered plies, i.e., 3, 5, 7, etc., so that the grain-direction of the outer layers each run the same way to counteract distortion known as *warping* or *cupping*. It is also necessary to have an equal distribution of ply-thicknesses on either side of a theoretical centreline (*neutral axis*). So, in the case of 3-ply, the face and backside veneers must be of equal thickness and the middle veneer (running in the opposite direction) must be twice the size of their combined thicknesses.

Pocket-screwing: This term equates to 'skew-nailing', when nails (as in the fixing of vertical partition-studs to horizontal head- and floor-plates) are driven in at an angle through the side of the stud, into the abutted surfaces of the floor- or head-plate. The difference is that pocket- screwing (apart from being done with screws instead of nails) is done via small countersunk *pockets* drilled or gouged into the vertical surfaces – usually of components such as the back-side surface of top riser-boards of a staircase, before the stair is fitted against a landing – or an upper floor's trimmed edge-joists.

Pointing: *Figure 59*: This term refers to the finished appearance of a wall's narrow (9mm/⅜ in. wide) mortar joints seen on face-brickwork or blockwork, etc. Traditionally, the mortar joints in newly-built exterior walls were raked-out to a depth of about 12mm/½ in. and filled (pointed) with a stronger mortar-mix to a ratio of 3:1 sand and cement, which was trowelled (pointed) to a smooth finish. If done correctly, a fatty laitance was brought to the surface of the pointing, giving it a smoother, closed-aggregate finish, which extended its lifespan from erosion. Modern-day pointing avoids the costly practice of raking-out and pointing later with a stronger mix – and *points* the weaker mortar used for bricklaying as the

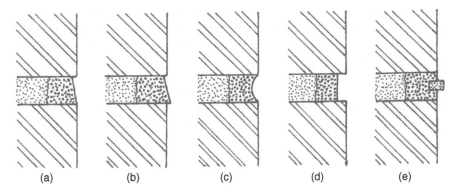

Figure 59 (a) Weather-struck pointing; (b) Weather-struck-and-cut pointing; (c) Curve-recessed (ironed-in) pointing; (d) Square-recessed pointing; (e) Protruding tuck pointing

work proceeds. This is done by *ironing* the partially-set mortar with a round-ended, reflex-shaped pointing tool (instead of a small-sized pointing trowel), to produce a concaved shape that was traditionally referred to as *curve-recessed (ironed-in) pointing*.

Other traditional forms of pointing (a few of which are still used) include: (a) Weather-struck pointing; (b) Weather-struck-and-cut pointing (leaning in at the top, but projecting to a cut line at the bottom); (c) Curve-recessed (ironed-in) pointing; (d) Square-recessed pointing; (e) Protruding tuck pointing. Note that **tuck pointing** is referred to separately in this book.

Portland cement: Ordinary building-cement made from a slurry mixture of clay and limestone; so-named historically because – when hardened via its chemical *setting action* – it was thought to resemble Portland stone.

Portland stone: A limestone from Dorset with a dull, grey colour, extensively used historically for elevational façades of large buildings in the city of London – presumably to pre-empt their eventual greyish degrade from the city's polluted atmosphere.

Post-and-panelled fence: This description usually refers to relatively lightweight posts of 75mm/3 in. x 75mm/3 in. section and preformed, lightweight panels of inter-woven (basket-weave style) thin boards of approx. 6mm/¼ in. thickness. Such panels, sandwiched and framed with light-weight edge-timbering, are fixed (nailed) through their edges to the inner, side edges of the posts. Although less expensive than other wooden fences, they are well-documented as having a limited lifespan.

P/P: This is a relatively modern reference to timber being ordered as *prepared*, replacing the traditional, more definitive abbreviation of **PAR** (*planed all round*).

Pressure-treated timber: This refers to a process of presale preservation-treatment of timber sections, such as floor joists and roof-tiling battens, etc. The timber sections are stacked into large, sealed cylindrical vats, which are then filled with patented preservative liquid. The liquid is then pressurized for a set period of time, to force it into the timber's outer cellular-surfaces.

Prime-cost sum: This refers to a guesstimated figure given in a *Bill of Quantities* to cover an area of unspecified work, or a component of unknown cost.

Professional Consultant's Certificate (PCC):   This important certificate is a transferable document that guarantees (for a maximum period of six years) that a named property, which has had building work carried out on it, has been subject to inspection by a consultant who is a member of an institute such as the Royal Institution of Chartered Surveyors (RICS) and that the property satisfactorily met the building control standards. Note that if a surveyor, on a rather time-limited Homebuyer's Report survey, has doubts about the quality of certain aspects of new-looking building work carried out on the property, he or she are now being advised to recommend that their client ask the vendor whether the work is covered by a PCC.

Professional Indemnity Insurance (PII):   Such insurance does not come cheaply, but is an essential facet of a practising surveyor's working life. It is required to protect him or her (or their firm) from any financial penalties that may be levied against them in court via litigation of being sued by a client (the purchaser), or the vendor, for alleged negligence or misinformation. Although surveyors cannot operate without such an indemnity against the likelihood of unaffordable financial penalties, it is a sobering thought to realize that involvement in such court actions can result in the surveyor (or their firm) having to pay increased PII premiums. Such a thought must be nurtured whilst surveying, to promote careful and accurate report-writing.

Progress charts: Time plays an important part in all building contracts and *completion dates* are usually agreed and written into the contract – with accruing financial penalties incurred unless the builder keeps a record of any justifiable holdups. These may be caused by weather on certain parts of the work, non-delivery of materials promised for agreed dates, or extra work created by the architect via *variation orders*, etc. So, detailed progress charts, or *critical path analysis charts* are commonly used to chart the estimated time needed for the various overlapping operations. This will include setting out and excavation, laying of the concrete foundations, brickwork up to DPC level, excavation for pipe-runs and drain-laying, etc. Each operation is shown graphically in relation to each other, with a starting date and an expected finishing date.

The Site Manager or Agent in charge of a contract should realize the importance of the chart and – if he or she is experienced – know how easy it is to lose time and wander off the path, which really is critical if and when this happens.

Property valuation: From a buildings' insurance point of view, there is a mathematical formula related to the approximate squared-metreage of a dwelling x the number of floors, whereby the resultant sum is multiplied by a fluctuating (usually annually-increasing) sum of money, to arrive at the rebuilding cost. But this does not account for the actual plot of land, which is usually quite valuable nowadays. And although this could be estimated and added to the sum (to arrive at a more realistic figure), it would still be a figure that has ignored the power of market forces.

Such a force dictates that the selling price of a property is governed by how much a buyer is prepared to pay for it, which – as Estate-Agent's valuers know – is always related to certain key factors; Location always comes top of that list, in relation to a property's overall desirability factor; its room-size and layout, the number of bedrooms, an appealing bathroom and a shower-room, a modern kitchen, a utility room, a garage, a decent driveway, the nearness of a shopping area, a good school, a park or wooded area, a hospital, etc. Not all of these things will be expected by everyone, of course – but any upgraded areas may make a property more valuable than the annual-percentage market-increases added to the original purchase price by Zoopla's excellent, but *blind* house valuations.

Protimized timber: See **Pressure-treated timber**

Provisional sum (pc): This refers to a *contingency sum* in a Bill of Quantities, to be set aside for the possibility of additional cost for unforeseen work, or for the unknown cost of an item.

Pseudo solid-walls: See **Snapped headers**

P-trap soil or waste pipes: See **S-traps** and **P-traps**

Pug: This term is used in certain parts of the UK in reference to mortar used for plastering or bricklaying.

Pugging: Historically, this referred to a crude form of soundproofing, fire-resisting, or thermal-insulating of timber-joisted floors and/or timber-stud walls. It was achieved by floor-joist voids being roughly boarded on their lower, inner edges to hold an infill of *pugging* in the form of a *'no-fines' concrete* (i.e., an aerated form of concrete, comprising ballast and cement, but lacking sand and fine aggregate); or, in the case of internal timber-stud partitions that were found to be packed with pugging (after the lath-and-plaster cladding had been disturbed/removed), this was achieved by an infill of very rough common-brickwork/*brick-nogging* between the intermediate studs and the diagonal bracing.

Pulley stiles: The essential, vertical side-members of a traditional boxframe window, in which the four, top-positioned pulley wheels are housed – two on each opposite side. The stiles are shallow-grooved centrally to house the parting beads, tongued or left square-edged (on cheap work) on their outer edges; and have an accessible pocket formed in each inner-face, to enable retrieval/insertion of the pulley-corded sash-weights.

Purlins: See *Figure 11* on page 22: Horizontal timber-beams in a traditional site-cut-and-pitched-roof, that run at right-angles across the undersides of the common rafters, usually at the midway point between the ridge and the wall plates. If the length of the common rafters does not exceed 2.440m/8 ft. from the wall-plate bearing to the ridge centre-line, then purlins are not usually required.

Purlin struts: See *Figure 11* on page 22: Traditional timber-*props* that are placed obliquely within a roof-space to support the longitudinal purlins that carry the mid-area load on the rafters. The bottom seat- and plumb-cuts on the struts must relate to *straining-piece plates* and a load-bearing wall.

Purpose-built flats/apartments: This self-explanatory description is only enlarged upon here to make the distinction between buildings that were and still are built purposely as blocks of flats, for separate-family occupation – and houses, that were built purposely as houses (for single-family occupation) – and to underline the fact that the latter, having been converted into flats, may be weak on soundproofing, etc., if they were converted prior to the amended Building Regulations in 1991 (described herein under **Flat-conversions**).

Purpose-made valley-tiles' valley: Plain tiles are used with these purpose-made concrete valley tiles, whose surface-shapes are very similar to large, upside-down *bonnet hip-tiles* – but, because of their concaved top-edge shape (opposite to a bonnet-tile's convex bottom-edge shape), they resemble a concaved heart-shape – with the heart-shaped point removed.

After the skeletal roof has been overlaid with a breathable, impervious membrane, such as Tyvec Supro or Marley Supro, and the tiling battens are fixed to the tiling gauge and mitred into the valley junction, a number of under-eaves tiles are laid and mitred into the return angle. The first concaved valley tile is then laid on, checked, adjusted and marked onto the two adjacent under-eaves tiles. These are then cut and repositioned and a number of first-course eaves' tiles are laid to-the-bond on either side of the right-angled junction. The valley tile is repositioned on top of these and its tapered sides are scratch-marked on to the two eaves' tiles below. When cut, they are then re-laid and nailed to the tiling battens on each side of the valley, thus allowing the first valley tile to be laid back into its final position.

This sequence of laying bonded tiles (tiles centred over the tiles below), marking and cutting-in the valley tiles, then fixing them and laying another valley tile, is continued up the valley until the valley is formed – prior to the main roof tiling.

Figure 60 Putlog scaffold tube

Putlog scaffold: *Figure 60*:   This temporary tubular-alloy scaffolding is also referred to as a bricklayer's scaffold. This is partly because it is initially built for bricklayers, but also because bricklayers sometimes build it themselves. It is erected in incremental stages of 1.52m/5 ft. *lifts*, as and when the newly-laid brickwork has reached that workable height – and has *set*/*hardened* enough (usually overnight) to withstand being loaded. The flattened, tubular ends of the putlogs (which equal a brick-joint's thickness) are laid on the outer-edges of the wall and are attached at their outer ends to the vertical *standards* and horizontal *ledgers* of the outer, diagonally braced tubular-framework. Note that, for safety's sake, this scaffolding needs to be secured by *outriggers* – scaffold tubes fixed to it at an angle to the ground, acting like *raking-shores*. Also note that, whilst being dismantled – not afterwards – the mortar-bed slots created by the removal of the putlogs, needs to be neatly *made good* with matching mortar. This making good is often done by the dismantling scaffolders, understandably, but the standard of their work can add annoying blemishes to a wall.

For comparison, see: **Independent scaffold**, written herein under that heading.

Putty: This is comprised of whiting (ground chalk) and linseed oil to make a plastic putty-mix, that is still used for bedding and face-bevelling single panes of glass in traditional-glazing to wooden casements and sash windows, etc.

PVA: Polyvinyl acetate adhesive; a common woodworking adhesive, usually for interior use.

Q

Quadrant: This usually refers to a small, quarter-round softwood beading, tradition-ally used as a cover-mould to conceal gaps caused by shrinkage/contraction of timber components.

Quango: A group of board members set up and funded by central government, to supervise or develop a particular activity (or regulation) related to the general public.

Quantity Surveyor: A technically qualified person, who is usually a member of a pro-fessional institute such as RICS (the Royal Institution of Chartered Surveyors) and/

or the CIOB (the Chartered Institute of Building), after having attained a technical education and qualifications in building-related subjects. They measure and quantify the detailed work shown on the architect's drawings and compile an A4-sized *Bill of Quantities* – copies of which will be sent to a number of chosen contractors, inviting them to competitively submit *tender*s (built-up prices) for the proposed work. During certain defined stages of the contract being carried out, Quantity Surveyors also measure the work that has been completed and are responsible for issuing signed payment certificates to the contractor.

Quarry tiles: Common, burnt clay, unglazed tiles, usually red- or buff-coloured, traditionally used for scullery (kitchen) floors and window-sills. Although unglazed, they are not porous.

Quarter-space landing: A relatively small stair-landing in the midst of a quarter-turn stair flight, that separates the direction of travel by a quarter of a circle, i.e., 90°.

Queen-post roof truss: See *Figure 55* on page 148: Historically, such timber trusses were used to form the habitable area of gable-ended or hip-ended mansard roofs. They provided the structural framework for the steeply-pitched outer surfaces of the lower parts of the roof – and shallow-height King-post roof trusses were superimposed upon them to form the structural framework for the shallow-pitched upper surfaces of the roof. Such roofs nowadays – still seen occasionally – are built using modern techniques, involving steel beams, taking their bearings from the gable walls.

Quirk: In woodworking terminology, this refers to the very narrow and very shallow (about 3mm/⅛ in. width and depth) groove that separates a moulded feature from the remaining width of the timber. Usually, such a quirk is related to a *staff-bead* mould.

Quoin: The outer corner of a wall.

Quotations: These should contain a clear, written description of the work to be carried out in relation to the final cost. If this excludes VAT, it should be *clearly* stated. To my mind, a quotation of '£2,800 including VAT' is ambiguous. Does it mean that the contractor has *included* it, or that the contractor wants you to *include* it? And because neither party can *exclude* VAT, surely it would be clearer to state '£2,800 *excluding VAT*'. There also seems to be a degree of trade and public confusion between the terms *Quotation* and *Estimate*, with the latter causing problems when either party believes it to be the bottom-line sum to be paid.

R

Radon gas and building sites: In the UK, this gas is found in the ground of some areas more than others. It is created when natural radioactive uranium decays slowly in the ground and seeps to the surface. And because of the way homes are heated and ventilated, a certain amount of this gas gets through a dwelling's ground floors. Therefore, ground-floor rooms are subject to most of the radon exposure. And if a dwelling is subject to high levels of radon over a long period of time, this can apparently lead to carcinogenic health-problems. But it seemingly depends on whether your dwelling is in an area likely to be subject to *high* levels of radon. Companies such as *Groundsure Location Intelligence* (info@groundsure.com) supply comprehensive details on individual property-locations. Note that this information on radon gas was gleaned from publichealthmatters.blog.gov.uk.

Raft foundations: See **Concrete-raft foundations**

Rainwater chute: This description refers to a waterproofed outlet through a parapet wall, that discharges rainwater from a tapered-, valley- or box-gutter into a hopper-head serving an external downpipe. The chute is usually two brick-courses high (150mm/6 in.) x a half-brick wide (112mm/4⅜ in.) and it is usually made of heavy-gauge lead.

Rainwater gutters: These u-shaped or semi-circular-shaped channels, as are commonly used at the edges (eaves) of roofs, are essential for carrying away rainwater that otherwise would run off the roof slopes and likely cascade down the wall-surfaces. Incidences of this nature – when gutters are faulty, blocked-up with moss or lack sufficient downpipes – are well known to cause penetrating-dampness to a building's walls. The gradient or *fall* of guttering should be 1 in 600. So, a 3m length of gutter would be 3,000 ÷ 600 = 5mm/¼ in. approximately. Such a slight fall is hardly noticeable on a short run, but on a long run within the relatively shallow depth of a fascia board's surface, it demands that the uppermost (highest) end-bracket be fixed as high as possible to the roof's overhanging tiles or slates. Then, to ensure a straight-line fall, the bracket at the opposite end of the run (nearest to the rainwater-outlet and downpipe), should be lowered according to the gradient ratio of 1 in 600. Once these two brackets are fixed, a taut string line is stretched and tied across their inverted-edges – and the mid-area brackets are then carefully fixed in relation to the line at not more than 1m/39 in. centres.

Modern-day eaves' gutters are predominantly of plastic (uPVC), as opposed to the traditional gutters of cast iron, etc. The sectional sizes and shapes vary, with the semi-circular shape (which is referred to as 'Half Round') being perhaps the most common. Other shapes are 'Square', 'Ogee' and 'Deep flow'. The latter has the advantage of coping quite well when there is a deficiency of rainwater downpipes and gullies – or the pitch of a particular roof is very steep.

Rainwater hopper-head: See **Hopper head**

Rainwater pipes (RWPs or downpipes):   Calculating the *gutter* and *rainwater-pipe* sizes for a known superficial area of pitched- or flat-roofing in the UK (other than Scotland), can be quite easily done by referring to Tables 1 and 2 in Section H3 of Approved Document H in The Building Regulations. They show six references to roof-area sizes ranging, from 6m^2 to 103m^2, with the recommended gutter-sizes (based on half-round gutters) ranging from 75mm to 150mm, and outlet/pipe-sizes ranging from 60mm to 90mm diameter (based on circular downpipes). Note that Approved Document H: Drainage and Waste Disposal makes many references to BS EN 12056-3: 2000, a European Standard.

The number of downpipes required for a particular roof-design and size is extremely important, but it seems to be elusively referred to in most reading matter as an architectural responsibility at the design stage – with nobody seemingly wanting to take responsibility for a shortage of downpipes, causing overflowing gutters. However, experience dictates that if there are too many right-angled, internal- and/or external-bends, there is more potential for overflowing during heavy rainfall. And the design-stage reference above to the number of downpipes, also relates to a correctly positioned number of rainwater gullies in the designed surface-drainage system below.

Rainwater shoe: A short right-angled spout at the foot of a downpipe/RWP, acting to direct the outflow away from the base of the building.

Raised-and-fielded panelled doors: *Figure 61*:   This technical description refers to traditional timber-panelled doors that were comprised of a number of softwood or hardwood panels, which had sunken, bevelled rebates formed around their four face-edges, leaving a flat, *raised field*, or an extremely-shallow *pyramidal raised field* as the central feature of the panels.

Raised (or Projecting) fire-break party-walls: *Figure 62*:   In old-style semi-detached or terraced houses (up until about 1950), the separating party-wall between the dwellings, usually of 225mm/9 in.-thick single-skin brickwork, was built-up between and beyond each property's timber roof-structure to act as a fire-break barrier. These raised projections, kept parallel to the sloping (pitched) roof-surfaces, protrude by between 300mm and 400mm. They are either capped with coping stones, bricks-on-edge with a *tiled-creasing* (double-layered and lapped quarry tiles) below, acting as a DPC, or the raised walls are rendered on their sides and top surfaces with sand-and-cement. Each party (the leaseholder or freeholder) shares the wall and is responsible for the maintenance and upkeep of their half, on each side of a theoretical centre line in mid-area of the wall. Note that any property displaying raised, fire-break party walls is also displaying potentially very vulnerable parts of the roof-fabric. Surveyors on a survey should give them close scrutiny with binoculars and – essentially, on the interior inspection – check the surfaces of the party wall with a moisture meter in the roof-void (loft).

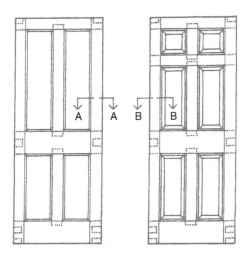

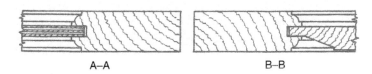

A–A   B–B

Figure 61 Elevations and sectional views through the stiles and panels of A-A, a plywood-panelled door - and B-B, a raised-and-fielded panelled door. Note that the inner edges of the stiles and rails are ovolo-moulded

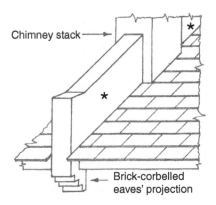

Figure 62 *Raised, fire-break party wall above the pitched roofs of terraced houses

Raking back: This term refers to the bricklaying practice of building up each opposite end of a wall before laying the bulk of the bricks in between. This is done by carefully forming step-bonded corners that – in the case of a stretcher-bond pattern – diminish by a half-brick length each course and rise by about 900mm/3ft. This practice (modified nowadays by using patented profile-ends) allows the corners to be plumbed carefully with a spirit level and checked for incremental height from fixed datum points at the base of each corner (quoin).

Raking out: See Pointing

Raking Shores: *Figure 63*: These traditional forms of *shoring*, which also include *flying shores* and ***dead shores*** (covered separately in this book), are designed to give temporary support to one or two buildings whilst structural repairs or demolition tasks are carried out. So, raking shores are thus required when a mid-terrace building is to be demolished, structurally repaired or altered. Basically, one or more raking shores (four being usual) are built against the front- and rear-elevational walls (two pairs on each elevation), laterally positioned as close as possible to the (unseen) party-wall front-and-rear-elevation junctions. This support is usually required until the subject building is either demolished, structurally altered or repaired. Such work may also involve *underpinning* the party walls.

A traditional raking shore consisted of **1)** a timber *wall piece* (*wall plate*), that was side-fixed vertically to the brickwork with wrought-iron wall hooks; **2)** a short, timber *sole plate* that was slightly angled into the ground (in a right-angled position opposite the plate on the wall), and at a distance from the subject wall that equalled about one-third of the highest raking-shore's height; and **3)** a number of heavy timber raking shores (one to support the centre of each floor-bearing in the wall) that fanned out upwards from their close-proximity on the sole plate, to press against the wall-plate in preformed housing-joints. Then small, square, projecting oak *needles* were driven into preformed mortise holes at the head of each raking shore (and through into the wall). Oak *cleats* were then fixed into preformed housings above each needle. Finally, timber-boarded side-bracing was fixed across the fan-shape at intervals.

Note that the basic principles of raking shores might nowadays be incorporated in the form of tubular-scaffolding arrangements, using a *minimum* of timber components but an *enormous amount* of tubular scaffolding and diagonal bracing.

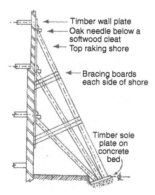

Timber wall plate
Oak needle below a
softwood cleat
Top raking shore

Bracing boards
each side of shore

Timber sole
plate on
concrete
bed

Figure 63 A traditional, timber 'raking shore' to a 3-storey building

Ramp: 1) A section of *wreathed* handrail, as part of a geometrical-stair balustrade, which is a concave shape on its top and is/was used to join a raking handrail to a higher, level handrail, via a mitred joint; **2**) A purpose-built floor-slope between different (internal or external) floor levels, specified in Approved Document K1, under Stairs, ladders and ramps.

Ranch-style fence: One particular arrangement of this consists of pre-mortised, pressurized preservative-treated posts and horizontal chestnut palings, usually forming three rows. The palings have extra-long, oblong-shaped tenons, machine-scalloped at their ends, which are left protruding after being fixed by nailing through the face- or rear-side face of the posts. Each post has six through mortises, two side-by-side for each row. This avoids the need for stub-tenons, makes it easy to adjust the posts to verticality before nailing through their face-sides to fix the tenons – and the protruding tenons seem to add to the fence's ranchiness.

Random-rubble walls: See **Rubble walls**

Rapid-hardening cement: Building cement to which an accelerator has been added at the manufacturing stage. When mixed with sharp or soft sand, such mortar gives a necessary advantage on live-drain repairs, etc.

Rat-trap bond: A hybrid brick-wall bond that creates a 215mm/8¾ in.-thick wall that is neither solid-, nor of cavity-construction. It is a mixture of both; and was done historically to economize on bricks. Although such walls were structurally sound, their susceptibility to lateral-damp penetration was questionable and they were superseded by present-day double-skin cavity walls. Rat-trap bond is recognizable facially by a Flemish-bond pattern of alternating headers and stretchers – but, the identifiable feature to look out for is that, *all of the bricks are laid on their edge*s. And although perhaps difficult to visualize, this on-edge arrangement creates a latticed framework of small, stepped cavities. Hence, these walls are partly solid and partly of cavity construction.

Rats, mice and squirrel intruders: See **Rodents in dwellings**

Ready-mixed, transportable concrete: In its initial fluid state, this ubiquitous product, which consists of a mixture of various ratios of **1**) sharp sand, **2**) aggregate/ballast (gravel of various-sized pebbles and rock fragments courser than sand), **3**) Portland cement and **4**) clean water, can be pre-ordered – mixed whilst en route to the site in the truck's revolving drum and off-loaded via chutes, or (if unavoidable) via prearranged pumping.

When ordering, the Ready-Mix companies usually want to know: **1**) The type of concrete required (whether prescribed or the regular design-mix); **2**) whether a minimum compressive strength is required; **3)** whether a type of cement is specified; **4)** whether a maximum size of aggregate is required; and **5)** the slump or workability of the mix, if being pumped.

Rebate: A recessed or inverted right-angled return, formed on the edge of timber, or other material – such as the rebated edges of door-linings (to accommodate a door).

Recessed pointing: See **Pointing**

Redwood and whitewood: These generic terms are sometimes used in industry to describe grades of *softwood*. Redwood refers to good quality softwood such as Scots pine, Douglas fir, red Baltic pine, etc., which has very close, easily discernible annual rings (denoting slow structural growth, usually a necessary ingredient for strength), a healthy golden-yellow or pinkish-yellow colour and a good weight – and *whitewood* usually refers to a poorer quality of softwood, such as fast-grown, low-grade European spruce (*picea abies* from the *Pinaceae* family), which has very wide (barely discernible) annual rings denoting fast growth and a lack of thick-walled, strength-giving cells (tracheids), a pallid, creamy-white colour and an undesirable light weight. It is not very durable and usually has hard glass-like knots, which have a tendency to splinter and break up when planed or sawn. However, as an exception to this description of whitewood, see the references to *whitewood* described under the heading of **Accoya® wood**.

Reeding: See *Figure 5* on page 9: This decorative feature, consisting of a number of closely-spaced, inverted semi-circular convex-shaped mouldings, also referred to as *infilled fluting*, can be found around the shaft of traditional, Ionic columns and on the face-side of rectangular-sectioned architrave, or other moulded work. When used on architraves, plinth blocks are used at the base and ornate cornice blocks replace the mitres at the head.

Regularized: This is a modern term, referring to a relatively new process applied to commonly-used first-fixing timbers – such as floor joists and stud partitioning – all being of more-precisely machined sizes and with rounded-edge arrises. Prior to this, sawn (unplaned) timber was used (the sizes and shapes of which were erratic) and the unplaned arrises often caused bad injuries from split-arris splinters.

Reinforced-bitumen roofing felt: Traditional non-breathable roofing felt, reinforced with embedded jute hessian, for laying over pitched rafters prior to fixing the spaced tile- or slate-battens. Note that roofs that have been underlined with non-breathable felt, ought to be ventilated at the eaves or in a number of designated mid-areas of the roof slopes.

Reinforced brick soldier-arches: See **Soldier arches**

Reinforced concrete: Interestingly – building-knowledge-wise – it should be realized that *pre-cast* or *cast-in-situ* concrete components, such as beams, floors (especially suspended floors), lintels over doorways and windows, window sills, etc., all have Achilles' heels in the form of a weakness in their lower, tensile-stress (tension) areas. The top areas of suspended concrete, being subjected to compressive-stress (compression) do not normally suffer this (unless their ends are built into walls), when, therefore, those limited, upper areas are also then subject to tensile stresses (tension) as well. Hence, the need for steel reinforcing bars to be buried in the tensile-stress areas, close to the concrete's bottom – and sometimes cranked- up and loop-ended near their constrained ends, to address the tensile-stress areas created close to the concrete's top.

Usually, reinforcement in pre-cast or cast-in-situ concrete components is designed by a structural engineer and can be quite complex, requiring steel-fixers to wire them up on site or in factories, if the components are pre-cast.

Relining damaged, underground drain-pipes: If traditional underground drainage pipes, of salt-glazed earthenware (SGW), get continuously blocked (usually via fine tree-roots seeking moisture through cracks/fractures in the pipes), there are three options: **1)** When blockages occur (detected by odours or foul water rising up within the inspection-chambers/manholes), call out a drain-clearing specialist; **2)** Have the pipe-run surveyed by CCTV equipment, with a view to excavating and renewing/replacing the drainage run, or **3)** Lessen the expense by having the SGW pipe lined with a glass fibre lining-tube.

Information received via *westerhamdrainage.co.uk* (in East Sussex), recommend the use of *RSM Lining Supplies Global Ltd*, who advise in their online advertisement that the so-called **drag-in lining** (or **pull-in-place lining**) is the usual method for relining small diameter (100 or 150mm) pipe-repairs, but only if there are two points of access via inspection chambers (manholes), one on each side of the pipe to be repaired by relining.

The process involves inserting and locating a long, interlaced-felt-material tube (that has been impregnated with a thermosetting resin), which is pulled/dragged through the pipe between one manhole and the other. The tube is then inflated by using a *calibration* hose. When the liner is cured (hardened) in situ, it forms a seamless, jointless repair to the pipe, effectively creating a new pipe within an existing pipeline. Finally, the entry and exit ends of a relined drain are advised by RSM to be tidied up by being sealed with epoxy putty.

Render and set: This can refer to ceilings, but usually refers to the plastering of wall-surfaces that are *undercoated* (rendered) with a *coarse* material and *set* (finished) with a fine *finishing* plaster.

Rendering: This is a reference to the plastering of internal or external wall surfaces, either with a traditional mixture of sand, cement and water (for exterior work), or modern materials such as one-coat multi-finish plaster for interior work – and ready-mixed, silicone-based coloured render for exterior work. The rendering and painting of exterior surfaces – arguably done for aesthetic or cost reasons – was traditionally referred to as *stucco*. Such treatments are sometimes questionably applied to degraded or defective face-brickwork, as a cheaper alternative to repairing and/ or repointing.

Repointing: Over a period of time, the *pointed* mortar-joints can be affected by the weather and start eroding and crumbling away. Usually, the walls facing a north-westerly direction are worse affected. Once the erosion has broken down the surface-laitance of the pointing and exposes the more-open aggregate of the mortar, penetrating moisture may enter the wall. At this stage, a judgement needs to be made regarding repointing. This will involve *raking out* the bed- (horizontal) and perp- (perpendicular) joints to a depth of at least 12mm/½ in. before repointing with a so-called *stiff* mortar-mix of soft sand and cement to a ratio of 3:1.

Restraint-strap additions: See *Figure 51* on page 124:   These galvanized-steel, right-angle-ended straps were originally introduced via the Building Regulations for fixing over timber wall-plates to prevent roof-structures from suffering wind-damage. Such action was prompted by changes in modern roof-design and the use of lighter-weight roofing members. However, after a very destructive storm in January 1987, when a great number of triangular-shaped gable-ended walls were seriously damaged/sucked out, wall-restraint straps for use on gable-end walls soon became a legislated requirement. So, hook-ended straps (at right-angles to the rafters) are now required to be built into the inner skins of the gable walls and a) screwed into the underside edges of the rafters and noggings at pitched-roof level, and b) screwed into the topside edges of the ceiling joists and noggings to anchor down the previously-vulnerable gable walls.

Restrictive covenants: These covenants (sometimes considered to be onerous nowadays) are written into the *Deeds of transfer* that originated from the building-developer of a property or properties – and they transfer from a vendor to the buyer once a purchasing contract of a property has been exchanged. Not all covenants are necessarily onerous, but they might be termed restrictive. For example, some covenants state that: (a) Radio aerials should not be fixed or seen on the front elevation of the building; (b) Washing lines or clothes' airers should be confined to the rear garden area; (c) No fences or walls of any kind should be built around the front, lawned areas of the property, adjacent to the public footpaths; (d) No mobile homes, caravans or the like should be parked or permanently sited on the front driveway or lawns of the property; (e) No building-extensions or outbuildings should be built at the rear of the property.

Retaining wall: Such walls are built to retain high-level ground on a sloping or split-level site. The hydrostatic pressures that build-up against such walls should be combatted by building-in a number of small-diameter (38mm to 65mm) weep-hole pipes near the base, usually at 0.6m., 0.9m., or 1.2m centres.

Return end: This term usually refers to the finished end of a length of fixed moulding, such as a plaster cornice, coving, or a timber dado rail, etc., that is *returned within itself*, i.e., shaped with the same mould-shape on its otherwise square end.

Return wall: This refers to a wall that continues around the corner, usually by 90°.

Reveal(s):   The relatively narrow *sides*, or return edges, of openings in walls for windows or doorways, etc. It seems likely that this term originated by being a reference to the amount of wall-thickness *revealed* before/or after a doorframe/lining or window frame was fixed in position. Note that when surveying a property, reveals (via windows or doorways) are very useful for *revealing* the thickness of a wall; thereby indicating whether they are of *solid* or of cavity-wall construction (according to their measurement).

RIBA: The Royal Institute of British Architects.

RICS: The Royal Institution of Chartered Surveyors.

Ridge: In roofing terminology, this refers to the apex of a roof.

Ridge board: The horizontally-positioned board at the apex of a traditional roof, acting as a spine, upon which the common rafters are fixed in opposite pairs.

Ridge tiles: These top-edge roof tiles can be of clay, concrete, slate or fibre-cement, etc., and of different sizes and shapes. Clay and concrete ridge tiles are usually semi-circular or segmental, but those used for slate and fibre-cement, etc., can be angular shaped.

Riding- or **Rider-shore:**   See *Figure 63* on page 200: When *raking shores* are used to provide temporary support to very tall buildings, with more than three upper floors, a short *back-shore* is sometimes laid on/attached to the back of the uppermost raking shore and a separate riding- or rider-shore (seated on folding wedges) uses the top-end of the back shore as its base (sole plate) and is fanned out to support a higher floor-bearing point or the bearing-point of the roof structure.

Right-angle forming methods: Such methods are usually only needed for initial setting out of foundation-trenches and (more importantly) the outer walls of the building, once the concrete foundations have been laid. This can be done with squaring instruments – especially lasers – or by accurate right-angled triangles, related to wooden (end-pointed) setting-out stakes driven to within 100mm/4 in. of the ground – and with ranging-lines. And, to tie the lines to the stakes, 63mm/2½ in., or 75mm/3 in. round-head wire nails are driven into their end grain and left protruding by about one-third.

The first pair of stakes, (1) and (2), should be set up to represent the *building line* (the face of the building), in excess of its width. Then two more stakes, (3) and (4), are driven in under the line, to determine the building's width via their nail-head centres. The next step is to create a *true right-angle*, either from stake (3) or stake (4). Once this is established, only the building's outer measurements are needed to create a parallelogram.

So, the right-angle can be formed by **a)** Pythagoras' theorem, where the square on the hypotenuse of a right-angled triangle is equal to the sum of the squares on the other two sides: $a^2 + b^2 = c^2$; or **b)** by triangulation known as the 3:4:5 method, whereby chosen units of measurement (say, increments of 300mm/1 ft. or 600mm/2 ft. or 900mm/3 ft., etc.) are used to form a right-angled triangle. For example, a triangle using 900mm units on its unsquared sides, would use 900mm x 3 = 2. 7m on one side, 900mm x 4 = 3. 6m on its opposite side – and then these sides would need to be adjusted (pivoted) to produce 900mm x 5 = 4.5m on the diagonal adjustment; or **c)** This is another method of triangulation that preceded Pythagoras' theorem, enabling right-angled triangles to be formed with integral sides, i.e., sides which can be measured in whole numbers or equal units. It is known as the 5:12:13 method. So, 5 units set up on its base line and 12 units on its adjacent side, would need to be adjusted to produce 13 units on the diagonal adjustment.

Note that to set up the 3:4:5 method of squaring from the nail-head centre of stake (3), being the front, left-hand corner (quoin) of the building, drive in stake (5) under the frontage-line to the right, with its nail-centre driven in to three chosen units. Next, ideally with the aid of two tape rules, loop one over the stake (3) nail-head and extend it at right-angles towards the rear of the site, then loop the other over the stake (5) nail head and extend it diagonally to the left to form an acute triangle with the first tape. Now, where the tapes intersect, move them one way or the other until your 4-unit measurement relates to your 5-unit measurement and drive-in stake (6) centrally below this point. Re-lay the acute-angled tapes over the stake and mark the exact intersecting point to enable the last right-angle forming nail to be driven in. If interested, see **Squaring methods used on site with laser levels** herein under that heading.

Rim latches and **locks:**   Basically, this description refers to either *traditional* face-fixing latches and locks or modern, more sophisticated, face-fixing locks that are usually accommodated in the face-fixing *door furniture*. The one remaining traditional type still to be seen ubiquitously on traditional entrance doors, is the *cylinder night latch*.

Ring-beam: A reinforced-concrete beam, at or near ground-level, that forms a continuous, rectangular-shaped beam foundation, usually on top of pre-driven piles.

Ring main: This refers to a continuous electrical power circuit, starting at the consumer unit and, eventually, after being connected to numerous socket outlets en route, returning to it.

Rise-and-going regulations (Stairs):   In a flight of stairs, the individual steps should all have the same *rise* and the same *going* to the regulated dimensions given for each category of stair in relation to the 2R+G formula – given here under **Stair design formula**.

Risers: This refers to the *riser boards* of a staircase, which traditionally were made of softwood or hardwood of 16mm to 18mm/⅝ in. to ¾ in. thickness. But, in the 1970s, 9mm to 12mm/⅜ in. to ½ in.-thick plywood was used quite effectively. Although, nowadays, this appears to have been mostly displaced by the use of similar thicknesses of MDF (medium density fibreboard), which is cheaper. But there is some evidence that plywood is still used on better-class work. Of course, on a hardwood staircase, the riser boards should be of solid hardwood – unless specified to be of hardwood-veneered plywood.

Rising-butt hinges or rising butts: These traditional, two-part hinges are useful when floors are out-of-level, or a door struggles to clear a fitted carpet. Each hinge is in two parts, each with a half-length projecting knuckle, which is at the bottom of the *frame-knuckle* and at the top of the *door-knuckle*. A round pin projects from the frame-knuckle, ready to be inserted in the door-knuckle. The knuckles are centrally joined via a helical spiral, thus allowing a door to rise up by about 10mm/⅜ in. when opened. Note that – to allow a door to close – the door's top, on the hinge-side, has to be planed off on the splay to about 6mm/¼ in. depth over about 225mm/9 in.
    Rising butt hinges are either left-handed or right-handed and to identify which hand is needed, name the hand as you face the door opening *away* from you.

Rising damp: See **DPC defects and remedial treatment**

Rising main (or rising pipe):   This refers to a point in a building where the water-/gas-/ or electric-supply rises up into a property from its final position underground. Note that, for water mains, British Standards (BS 4118) prefers the term *rising pipe*, instead of *rising main.*

Rod: 1) This term, which has many dictionary meanings, is used in the building industry to refer to either a long timber batten, to be accurately marked with the on-site storey-height of floor-to-floor levels (a traditional method for gaining precise heights from one floor-level to the next, for setting-out stairs and landings); or for subdividing the storey-height into brick-rise marks – thereby enabling a bricklayer to rest the **storey rod** on a *datum batten*, fixed to the wall at ground-floor level, to check the floor-to-floor incremental rise of his brickwork.
    2) A plain, or white emulsion-painted board, upon which full-size sectional-view drawings are made of items of joinery such as windows and doors, etc., showing the precise positions of their joints and shoulders. This detailed board is also referred to as a rod. In fact, in this second definition, the term rod accurately describes the abbreviated form of *r*eading *of d*rawings.

Rodding: This usually refers to cleaning out blocked drains with *drain rods*, which is a hazardous process, often making the problem worse by rods becoming unscrewed and irretrievably *lost* in the pipework. Nowadays, it is more common to clear blocked drains with power-jet hosepipes, which is a job for the professional drain-specialists – not plumbers.

Rodding-eye: See *Figure 52* on page 142: Some trapped gullies have *rodding-eyes,* which gives rodding-access to a trap-concealed drainage pipe, by providing a short pipe-run above the inaccessible trap. This short pipe-run access has its own screwed-down cover close to the trapped gully.

Rodents in dwellings: Whilst inspecting cupboards, roof-voids/lofts and accessible sub-terranean areas of a dwelling, one should look out for evidence of gnawing-damage caused by intruders such as rats, mice and squirrels. Open-topped cavity walls and small holes/gaps (usually via lack of maintenance or shoddy building) can often be found at the roof's eaves (and these can be gnawed to an entry-size by rodents); and, at ground level, broken air-vent grilles and poor masonry-infill around soil pipes leaving the building, can create rodent-entry points. Rats, apparently, having entered an under-ground drainage system, can climb up the cylindrical interior surface of a 110mm/4¼ in. plastic soil pipe (via their sharp-pointed claws and back-support against the pipe) and appear in the shallow, trapped water of a WC pan – or in the dwelling, if the WC's lid is left open. Special sewer-exit flaps are obtainable and can be fitted to inhibit the ingress of rats.

In accessible roof-voids/lofts, evidence of rodents can be detected by disturbance of loft-insulation (being used for nesting) – and plastic-sheathed, electrical cables and plastic pipes should be checked for signs of gnawing damage. Note also that rats leave traces of greasy, black marks on the surfaces of objects that they squeeze past/rub up against, that leaves an unpleasant smell. They, like other rodents, also leave detectible evidence of excrement, in the form of varying-sized, brown-coloured ovoid-pellets.

Roof cladding: See **Roof tiles** and **slates**

Roofing felt: **1)** An impervious, flexible membrane laid over the rafters of pitched roofs before the slate- or tile-battens are fixed, as a precaution against descending-damp via rain being wind-driven under the bottom- and side-edges of the roof-cladding (tiles or slates), or damp-penetration via capillary attraction and other leakages. This practice only started in the early 1930s. Therefore, any properties built prior to about 1933 that have not been retiled or re-slated since then, are likely to be unfelted and susceptible to penetrating damp, via broken slates or tiles or wind-driven rain. However, the roofs of some older properties have been **torched**, in an attempt to keep out wind-driven rain – but this technique failed over time via capillary attraction causing wet-rot decay to the tile- or slate-battens. Note that the technique of **torching** is fully described elsewhere in this book.

Note also that since the early 2000s, *impervious roofing-felt* has been superseded by polyethylene **breathable-*and-impervious*** roofing membranes, such as Tyvec Supro, Marley Supro, etc., negating the need for separate roof-void ventilation at the eaves.

2) Bituminous roofing felt (without hessian-mesh reinforcement) is also used on traditional, cold-deck flat roofs, in built-up layers (usually three), bonded together with hot coal-tar pitch. The top layer was traditionally finished with pitch-bonded spa chippings, but in recent years, a mineral-finished top layer has been used. These top coatings add a degree of protection to these relatively soft roof-surfaces and may increase their lifespan to a textbook expectancy of 10 to 15 years. However, there are thicker, mineral-felts nowadays that are torched on (with gas-filled blow lamps) to the surfaces of the felt underlays – and these have a guaranteed lifespan of 20 to 25 years.

Roofing-mortar: Site-mixed mortar for hip-, ridge- and valley-tile bedding was traditionally comprised of fine-graded sharp-sand and Portland cement to a mixed ratio of 3:1. This produced a very strong, weather-resistant mix which proved to be long-lasting, but it lacked a degree of easy-workability and instant-adhesion to the substrate – thus slowing down the site-work. So, over the years, uncontrolled site practices seem to have developed along a self-serving route of substituting bricklaying mortar (with a 6:1 ratio of *soft* sand to cement) for roofing-mortar. This is much easier to use, of course, with good workability and instant adhesion – but it is far less weather-resistant and much more susceptible to erosion and thermal cracking.

Note that publications by well-established authorities, such as the NHBC (National House-Building Council), have recommended the use of *sharp* sand in roofing-mortar, instead of *soft* sand. However, from a workability point of view, roofing-mortar mixes to a ratio of 1½:1½:1 (1½ sharp sand; 1½ soft sand; 1 cement) seem now to be an accepted compromise.

Roofing nails (for fixing tiles or slates, etc.):   The British Standards Code of Practice (BSCP) states that ordinary steel or galvanized-steel nails should not be used. The recommended list is between copper, aluminium, silicon bronze and stainless steel.

Roofing shingles: See **Wood shingles**

Roofing-square instrument: Even though on-site cut-and-pitched roofing has mostly given way to factory-produced trussed-rafter assemblies, there are roofing aids available for finding and marking out the various angles/bevels and lengths that are still required for certain parts of the factory-supplied assemblies (as well as for a much-reduced number of wholly cut-and-pitched site-assemblies). One of these roofing aids is a steel, metric roofing square, of 3mm/⅛ in. thickness, a 50mm/2 in. wide *blade* of 620mm/ 24½ in. length and a so-called *tongue* of 40mm/1½ in. width x 450mm/17¾ in. length.

The simplest way to understand it, is to think of its right-angled, outer-edges as metric scale-rules, with a 1:10 (1mm=10mm) metric scale. As an example, if the span of a pitched roof is 6.2m, it must be realized that this needs to be halved to 3.1m (because the isosceles roof-shape is comprised of *two* right-angled triangles and you only need to deal with *one*. This realization makes it a lot easier, as only one right-angled triangle (represented by the roofing square) needs to be dealt with to find the two essential angles and lengths-of-rafters required. The angle at the triangle's top equals the required **plumb-cut** angle and the other being at the triangle's base, equals the **seat-cut** angle. The hypotenuse of the triangle depicts the required *scaled length* of the common rafters. Finally, by this roofing method, the vertical height of the triangle equals the **rise** of the roof and its base is referred to as the *run* of the roof. The latter, as previously stated, being/meaning half the roof's span.

Finally, by using the protractor-facility etched on the square's tongue, (or, by using a simple, semi-circular protractor), the designer's roof-pitch (say, 38°) bevel (representing the roof's seat cut) can be marked on a timber straight edge. Then the steel-square's outer blade-edge is laid up against it and slid along the line until the scaled-run of the roof (3.1m ÷ 10 = 310mm) is registered. Next, whatever measurement has been produced (by the sliding action) on the outer-edge of the tongue, must be noted. This is the scaled rise, and, finally, the hypotenuse this produces is measured to give the initial scaled rafter-length.

So, 310mm run on the blade at a 38° pitch-angle, produces a rise of 239mm on the tongue and (more importantly) a scaled *rafter-length* on the hypotenuse of 392mm (3.92m). This dimension depicts the so-called *notional pitch line* from the outer corner of the wall plate, to the centre of the ridge board, to which an eaves' projection must be added and half of the ridge-board thickness deducted.

Roofmaster Instrument: This revolutionary roofing *square* (which is actually a relatively small, *right-angled triangle*) was invented by Kevin Hodger, a lecturer in Hastings, East Sussex, UK. It is an excellent alternative to a traditional **roofing square** and is a compact, precision instrument, measuring 335mm/13¼ in. on each right-angled side, and is of anodized aluminium with easy-to-read laser-etched markings. It gives angle cuts for all roof members *and* the lengths of rafters, without the need for separate complex tables. It is designed for easy use, whereby only the roof pitch angle is required to obtain all other angles and lengths. Surveyors requiring to know the pitch angle of an existing roof (to question, perhaps, the validity of the tile-type used on a retiled roof), can determine this easily within the roof-void. Roofmaster Squares are obtainable via kingsviewoptical.com.

Roof spread: See **Structural appraisal of low-rise buildings**

Roof tiles and slates: 1) Traditional **plain tiles** are either made of clay or concrete. The former is the more expensive, but are reckoned to have a better, softer appearance. Plain tiles are spoken of as having flat surfaces, but in fact, they are slightly cambered across their narrow width. They are approx. 265mm/10½ in. long x 165mm/6½ in. wide and approx. 12mm/ ½ in. to 16mm/⅝ in. thick. They are hung on the tiling battens by their two top nibs and can be fixed via two top nail-holes. Their much-revered weathering-protection arises from the fact that each row of tiles overlaps two courses of tiles previously laid below. Which means that when a number of bottom portions of tile break off over time, the two remaining layers below still give a good degree of weathering protection. Their lifespan is reckoned to be in the region of up to 80 years.

2) Concrete interlocking tiles: These tiles, because of their relative cheapness and greater covering-capacity with less tiles, have grown in popularity since the mid-20th century. The most popular of this type seem to be the *Redland 49*, identifiable by their flat-topped channels that give them their surface-strength and guide the rainwater down to the gutters. They are much larger (and heavier) than the *plain* concrete-tiles described above. They are approx. 380mm/15 in. long x 229mm/9 in. wide x 13mm/½ in. thick. They each have an interlocking side-lap of 25mm/1 in. width. They are hung on the tiling battens by their two top-end nibs and can be fixed via a top, central nail hole. Unlike plain tiles described above, concrete interlocking tiles only overlap the top edge of *one* tile below and because of this, they are referred to as *single-lap* tiles. Their life-span is reckoned to be in the region of up to 70 years.

(3) Slates: This natural, quarried material is converted into slates in a range of various sizes and thicknesses, from 407mm/16 in. long x 229mm/9 in. wide and with varying thicknesses ranging from 3mm/⅛ in. to 10mm/⅜ in. They have a good track record for longevity, but are expensive by comparison with other options, including artificial (fibre-cement) slates, although the latter are not yet known for longevity.

Roof trusses: Different forms of roof trusses, related to their spans and where they were used, have evolved over the centuries, from so-called **Mansard** roof trusses (with spans up to 5.5m/18 ft.); **King-post** roof trusses (with spans up to 6.1m/20 ft.); **hammer-beam** roof trusses (with spans up to 7.62m/25ft; **queen-post** roof trusses (with spans up to 10.67m/35ft; **Collar-beam** (open roof) trusses (with spans up to 7.20m/23 ft.6 in.); **Belfast** roof trusses (with spans up to 15.25m/50 ft.); **TRADA** (Timber Research & Development Association) roof trusses for houses (with spans up to 6.15m/20 ft.) in vogue in the post-WWII period. Nowadays, most newbuild dwelling-house roofs are wholly comprised of factory-designed and produced triangulated frames referred to as *trussed rafters* (trussed-rafter assemblies).

Roof windows: *Figure 64:*   This generic term, believed to have been originated by Velux in the late 1900s, refers to a more sophisticated and reliable version of traditional *skylight* windows. Seemingly regardless of the present-day number of different manufacturers

Figure 64 Roof window (skylight)

of *roof windows*, they all still seem to be referred to in the building industry as Velux windows. They are made from preservative-impregnated Swedish pine and clad on their exterior surfaces with sheet aluminium. The windows are of the horizontal pivot type, with patent espagnolette locks, seals and draught excluders and double-glazed sealed units. The windows are suitable for roof pitches between 20^0 and 85^0.

RSJ: This abbreviation is commonly used to refer to a ***rolled-steel joist*** (a ***steel beam***).

Rubber-clad roofs: This sheet-rubber roofing is technically known as EDPM (ethylene propylene diene monomer), which is a rubberized material, impervious to moisture. But it is still breathable, meaning that interstitial condensation (often trapped between differing layers of material) can escape to the exterior air. This roof cladding is very resistant to damage – and its advantages over other flat-roofing materials are that only a single sheet is required to be glued to the decking, before the perimeter edges are trimmed. If laid by a specialist, this roofing material is usually reckoned to have at least a 25-year lifespan.

Rubble: See **Back-filling** or **back-fill**

Rubble walls: These are built of natural, hewn stone and are either laid in lined courses or, more commonly, in irregular, unlined courses. The mortar joints are usually about 25mm/1 in. thick. On ashlar stone walls, the mortar joints are usually much thinner, up to 9mm/3/8 in. thick.

Rustic (or rusticated) ashlar: This refers to ashlar stone-block walling, on which the faces of the blocks have been left roughly- or symmetrically-hewn – but the jointing face-sides have been cut back with bevelled- and/or recessed-edges.

Rustic bricks: This description can refer to *sand-faced* facing-brick*s*, but usually refers to facing bricks with a zigzag-lined indented pattern on their face-sides and face-ends.

S

Saddle board: This carpentry term refers to a purpose-made triangular board, acting as an apex gusset-plate, which is fixed at the end of the ridge board and to the faces of the first pair of common rafters. It supports – and distributes the load of – the hip rafters and the crown rafter of a hipped-end roof.

Saddle piece: A roughly square-shaped lead flashing that initially covers (is dressed down to) various junctions between two or more sloping roof-intersections, prior to being overlaid with mitred tiles or slates.

Safe: See **Lead safes (for chimneys)**

Safety glass (in doors and windows, etc.): This either refers to glass of 6mm/¼ in. thickness, with a thin wire mesh embedded in it, formerly known as GWP (Georgian Wired Plate glass), or armour-plated toughened glass, etc. Note that clear GWP is usually used in the glazed aperture of FR (fire-resisting) doors, to enable visual appraisal of the other side during a fire.

Safety issues/violations: During a Building Survey, a surveyor should note and report upon such issues as **1)** the non-existence of handrails to interior stairs or exterior steps; the non-existence/removal of balustrading from stairs; **2)** stepped floor-levels/single steps, found occasionally in rooms or hallways – usually in old properties – that are not highlighted with fluorescent-coated, angled trim; **3)** dangerously irregular riser-heights of masonry/paved steps (seemingly DIY-formed), to exterior entrance doorways; **4)** non-safety glass of yesteryear, found frequently in glazed doors and windows below 800mm/31½ in. from a dwelling's finished floor level (ffl). If in doubt, look for the small, British Kitemark, or an EN number in a corner of the glass.

Salt-glazed ware (SGW): This term refers to the *glazing* of traditional earthenware drain pipes, that was needed to seal the earthenware's porosity – which was achieved by depositing salt into the heated kilns.

Sand-and-cement fillets: On old-build properties, one quite frequently sees sand-and-cement fillets (shaped like 45⁰ fence arris-rails) running along roof-abutments to walls. This is a cheap, inferior weathered abutment, used as an alternative to a lead flashing and soakers. Thermal expansion and contraction invariably crack the abutment over time, which can cause descending damp to enter the building's interior.

Sapwood: This refers to the active growth of a tree, that encircles the mature central heartwood to a greater or lesser extent, according to the species. Because of the sapwood's open grain and its abundance of sap and mineral content – even after seasoning – it has a lower durability than the heartwood and is therefore less stable.

Sarking: See *Figure 29* on page 68: Close-fitting square-edged sawn boarding, traditionally of 19mm/¾ in. thickness, or sheet material, fixed to the tops of rafters as a strengthening and insulating membrane, in addition to the sarking felt or breathable-membrane underlay laid upon it. Not seemingly used that often south of the Scottish borders. Note: For more information on this subject, see **Counter-battens** covered here.

Sarking felt or **Polyethylene Breathable-membrane underlay:**   As inferred above, this is draped over the rafters, with recommended edge-lapping of 150mm/6 in. The sarking felt is usually of bituminous flax and can still be used if the roof-void is well-ventilated – otherwise unvented roofs should be overlaid with breathable membranes such as Tyvec Supro or Marley Supro, etc.

Sash-balances: These spring-loaded, adjustable balances have been marketed for a number of decades now, as an alternative method of balancing vertically-sliding sash windows. They could be used to convert traditional boxframe window-sashes, that operate on pulley wheels, sash cords and cast-iron weights or they can be used when vertically-sliding sashes are to be fitted to a window frame with solid timber jambs (as opposed to the hollow, boxed jambs).

Sash bars: See **Glazing bars**

Sash cord: This purpose-made cord, available waxed, if preferred, is about 9mm/⅜ in. diameter and is sold in *hanks*. As implied, it is used for hanging sash-windows or rehanging them, if their cords have broken. One hank is usually enough to re-cord a pair of standard-size sash windows.

Sash fasteners: Claw-levered or pivoted-bolt fittings fixed to the meeting rails of vertically-sliding sashes, that can be levered or bolted to the receiving fastener on the aligned rail.

Sash lifts/handles: Traditionally, two D-shaped handles were fixed to the underside of the top sash's meeting rail (for lowering the top sash) and two metal-plate hooks (sash-lifts) were fixed to the inner-face of the bottom sash-rail (for raising/opening the bottom sash).

Sash pulley-wheels: These components (of various metals, sizes and quality) are mortised into the top areas of a boxframe window's pulley stiles (close to the pulley-stile head), to enable sash-cords to run over their wheels – after the cords have been fixed to the grooved-sides of the sashes and attached to the box-concealed, balanced sash-weights.

Sash weights: Made of lead or cast iron, for balancing the sliding sashes in a traditional boxframe window. They are in two balanced pairs, separated by **wagtails** housed in the boxed jambs of the windows. The sash cords are attached to them via a pre-set metal screw-eye in the lead and via a pre-cast countersunk top- and side-hole near the cast-iron top.

Sash windows: See *Figure 13* on page 30: This description refers to a pair of glazed, sliding sashes that consist of two sash stiles on each side, a sash rail at the head, two splayed meeting-rails for the mid-area connection of the two sashes, and a bottom rail that is at least splayed to fit a weathered sill. Note that, traditionally, pairs of top-and-bottom sashes were set out by manufacturers (or bench joiners) to receive equal heights of glass – which meant taking into account the fact that the bottom rail was about twice the depth of the top rail – and that the bottom meeting-rail of the top sash was rebated for glazing, as opposed to the top meeting-rail of the bottom sash that was grooved for glazing. Although this sounds complex, when being set out on a full-size sectional drawing (which, in trade-terms, is a *joinery setting-out rod*), it is relatively simple.

Schedule of Condition or Dilapidations Report: When a leasehold property is in the initial stages of being leased by a landlord to a tenant or tenants, sometimes one or the other of the two parties may decide (or in some cases, should decide) to raise a Schedule of Condition, via a Building Surveyor. If the property happens to be in *a good condition* (structurally and decoratively), such a schedule would/should convey this, and might eventually recompense the landlord for any damage to the property at the end of the lease, but if the property happens to be in *a poor condition*, such a schedule would/should protect the tenant(s) at the end of the lease from having to return the property in an arguably *better* condition.

Schedule of Works: This is like a Bill of Quantities, but on a smaller scale. It is written by a surveyor for a client who requires an appraisal/description of repair, maintenance work, small alterations or conversion to be done. This enables a client to present a technically-worded Schedule of Works to a builder, with the expectation of receiving a quotation, rather than an estimate – and a resultant job that meets the stipulated requirements of the surveyor and the client.

Screed: This usually refers to a layer of sand-and-cement laid over a concrete-floor to give it a smooth and level surface. Sharp sand and Portland cement to a 3:1 mix, should be not less than 50mm/2 in. thick. It should not be laid too wet, nor too dry, thereby allowing the final trowel-finish to bring the laitance to the surface.

Screw-gauge: For most screw-fixings, one needs to know the gauge-size and length of screws required. Gauge refers to the diameter of the shank in millimetres, which usually dictates the diameter of the head and the screw's length, i.e., the longer the screw, the larger the gauge must be structurally. So, screw-heads, which are usually formed at an angle of 45⁰ to the shank, end up being approximately twice the diameter of the shank. And woodworkers often need to know this head-size, to ensure that screws will fit correctly in countersunk holes in ironmongery such as butt hinges, where the flat-headed screws should sit flush or slightly below the surface. Therefore, to determine the diameter of a screw's head for certain jobs, one can either measure the actual diameter of the countersunk hole, or double-up on the screw-manufacturer's given shank-gauge. For example, a screw-box label describing 4.0 x 40mm CSK screws is calculated as 4.0 x 2 = 8mm diameter (approx.) screw-head size.

Traditionally (pre-metrication), screw-gauges related to imperial measurements were determined by measuring across a screw's head (or the countersunk hole) in sixteenths of an inch, doubling the resultant figure and subtracting two. For example, a head or a countersunk hole measuring five sixteenths of an inch = 5 x 2 = 10 minus 2 = 8 (a No. 8 gauge screw).

Screw types: Screws are mostly sold in cartons/boxes of 100 and 200 and are referred to by the amount required, their length, gauge, head-type, metal-type/finish and their driving-slot type, usually in that order – although screw-suppliers put the gauge before the length. The most commonly used head-types used in the building trades are (a) flat, countersunk (CSK) head, (b) the double-pitched countersunk head, (c) the bugle countersunk head, (d) the raised (RSD) countersunk head, and (e) the round (RND) head.

The raised or rounded part of the head is not included in the stated length. The lengths of screws generally range from 9mm/⅜ in. to 150mm/6 in., but only relatively short screws are available in RSD- and RND-headed types. CSK screws predominate for their structural use; and RND-headed screws are mostly used for fixing ironmongery.

The different types of metal and treated-finishes used (either for strength, cost, visual or anti-corrosion reasons, include: steel (S), galvanized steel (G), stainless steel (SS), bright zinc-plated (BZP), zinc and yellow passivated (YZP), brass (B), sherardized (SH), aluminium alloy (A), electro-brassed (EB), bronze (BR), black phosphate (BP) and black japanned (J). Most of the abbreviations listed here are used in suppliers' catalogues and sometimes on the manufacturer's screw-box labels.

Scribed joints: These, instead of mitres, are – or should be – used when moulded skirting-boards, etc., meet at internal right- or obtuse-angles. Whilst one end of a moulded board abuts the wall, the other board is mitred to produce a geometrical view of the skirting's exact profile. The profiled shape is then removed with the aid of a coping saw (or jig saw) and, if done well, should fit against the square-ended, moulded skirting member and should look like a mitre. One of the reasons for scribing skirting at internal-angle junctions, as opposed to mitring, is that scribing allows for the unavoidable lateral movement of the scribed piece when being fixed to the wall – and thereby produces a better *mitred-looking* joint.

Scrim/scrimming: Traditionally, hessian-woven scrim, in rolls of 100mm/4 in. width, were plastered over the abutment-joints of plasterboard to ceilings and stud partitions, as a form of reinforcement prior to *skimming* the whole ceiling or partition with a coat or coats of finishing plaster. Nowadays, the method is similar, but the scrim has changed to being made of self-adhesive woven-nylon in narrower rolls of 50mm/2 in. wide.

Scrolled handrailing: See **Handrail scroll**

Sealants: This term usually refers to the wide variety of sealing compounds available for use nowadays in simple, sealant-cartridges that fit simplistic cartridge-guns. This is a great improvement on the Cellophane-wrapped tubes of grey mastic of yesteryear, that had to be tediously and carefully fitted into the early models of cartridge-guns. The need for sealants is **1)** partly to address the differential movement and cracking that occurs between different materials up against each other (such as uPVC window-jambs up against brick side-reveals, **2)** partly to hide/cover any tolerance gaps that understand-ably cannot be avoided and **3)** to create a seal against moisture penetration.

The long list of sealants available includes **1)** Silicone sealant; **2)** Siliconized acrylic sealant; **3)** Multi-purpose silicone sealant; **4)** External frame sealant; **5)** Intumescent acrylic sealant; **6)** Lead-compound sealant; **7)** Roof and gutter sealant; **8)** Glazing sealant; **9)** Acoustic sealant/adhesive, etc.

Note that sealants were not used externally around windows and doorframes fitted before the mid-20th century. Instead, the abutment gaps between the head and sides of boxframe windows and doorframes, up against the brick side-reveals and the lintel or arch, were usually tediously pointed with *weather-struck-and-cut* mortar joints. If the abutment gaps were too wide, they were stuffed with screwed-up newspaper prior to being pointed. A glance around a window or entrance door of many neglected pre-WWII properties will often confirm this. Of course, such extra-large gaps nowadays would be initially filled with expanding foam from a cartridge-gun.

Seasoning: See **Air-seasoning of timber** or **Kiln-seasoning of timber**

Seat-cut: This is usually a reference to a horizontal sawcut made on rafters, such being one of the two cuts required to form a so-called *birdsmouth* – the other is a vertical *plumb-cut*.

Secondary (double- or triple-) glazing: This term usually refers to two separate windows fitted within the reveals (sides) of a window opening, regardless of whether the outer windows are single- or double-glazed. The second window, fitted on the inner edges of the side-reveals, is sometimes made in the form of sliding glass panels. Note, however, that one-piece triple-glazed sealed units are available nowadays.

Second fixing: See **First-, second- and third-fixings**

Secret gutters: These can be found on slated roofs, either where the end of a pitched roof abuts a wall (as with a stepped party-wall abutment), or – more likely – below the mitred slates of the inverted dihedral-angle of a valley junction. The former has a lead-lined box-gutter up against the wall, linked up to stepped lead-flashing and secreted by the oversailing slates – and the secret valley-gutter is comprised of two 225mm/9 in. x 25mm/1 in. sawn boards (forming the inverted dihedral angle), with tilting fillets fixed to their outer edges. The whole of this formation is then lined with lead – especially up-and-over the two tilting fillets – ready for the oversailing mitred-slates to complete the formation of the secret gutter. The only secret not well kept about such guttering, is that in leafy areas they have a tendency to block, which can cause descending-damp problems.

Secret-nailed flooring: This is not a secret to carpenters, who know that the nails – so-called *lost-head oval nails*, or *lost-head wire nails* – are secreted below the surface by being driven into the recessed edges of the top-splayed tongues that project on the tongued side of the boards. Unlike ordinary tongue-and-grooved boarding, which has square T&G edges and parallel tongues, the edge-detail for secret-nailed flooring differs in two design features: **1)** The edges are not square; the top-surface edge is further in than the underside edge. This is to give the underside edge more projection to receive the slightly-angled nails without splitting; **2)** The wider top-tongues are splayed by about 12° to minimize the risk of them being broken when the nails are driven and nail-punched in at an angle of about 60°.

Segmental arches: Brick- or stone-arches are still used to some extent over doorways and windows, if only acting as lintels – although many modern-day facework-lintels are secreted behind the brickwork. Geometrically, segmental arches are any size of segment less than a semi-circle.

Semi-circular arches: Geometrically, these are as their description implies (half a circle); brick- or stone-work used to create this shape is usually in the form of tapered *voussoirs.*

Semi-detached houses: This self-explanatory description refers to two conjoined houses that are joined by a shared party-wall – half of which (via an imaginary centreline), each owner is legally responsible for, regarding its upkeep and repair.

Semi-elliptical arches: True semi-elliptical shapes were not used for brick- or stone-arches, because the methods of setting out did not give a bricklayer or stonemason the necessary centre points as a reference to the radiating geometric-normalcy of the voussoir joints. So, *approximate* semi-elliptical arches, known as *three-centre-* and *five-centre-methods* were used. The three-centre method being most used, as the five-centre method is quite complex.

S-ended wall-tie bars: See **Wall-tie bars**

Septic tanks: These are sewage-collection and disposal-units, installed within the boundaries of properties in remote, rural areas, where public sewers are not available. Older-generation tanks, built below ground, are comprised of concrete-slab foundations, brick-built walls and reinforced concrete top-slabs with two cast iron inspection/access covers. The piped house-drainage discharges through the wall into the tank on one side, and (after the liquid effluent has passed under a central scum-baffle) it exits slowly into an outlet pipe in the wall on the other side. This pipe leads to a subsoil drainage/irrigation system that allows the soiled water to trickle into the ground. Great care has to be taken not to flood the ground (via inadequate irrigation trenching), because the soiled water contains dangerous bacteria. The remaining sludge and faecal solids that remain at the bottom of the tank – which will have been mostly decomposed by anaerobic bacterial activity – must be mechanically sucked out annually by specialist contractors.

Updated septic tanks nowadays – such as a ***Klargester*** – are made of GRP (fibre-glass) and are much securer treatment-plants than the traditional tanks described above. Once buried in the ground, only their ground-level inspection-frame and cover remain visible at ground-surface level. However, mechanical sucking out is still required periodically.

Setting coat: Traditional 3-coat plaster is referred to as **float, render and set**. So, *set* or *setting* refers to the finishing coat. The *floating and rendering* refers to the undercoats, which consisted of coarser material of sand, cement and (traditionally) lime. Then patented plaster, such as Carlite Browning, Carlite Bonding and Carlite Finish replaced traditional plasters – and these have now been superseded by so-called One-Coat plasters. And all of the above have now been largely superseded by plasterboard dry-linings.

Setting out: This refers to **1)** Driving wooden, pointed *pegs*, of about 600mm/2 ft. length x 50mm/2 in. square section, into the ground of a prepared building-site, to set out a building or buildings; **2)** Setting out full-size sectional, detailed drawings of items of joinery, such as horizontal sections and vertical sections of windows, or doors, as a construction-aid; **3)** Marking timber floors with chalk-lines to set out internal stud-partitions or marking concrete floors with trowel-marked mortar-spots to set out block-work partitions.

Settlement appraisal (internally and externally): Building Surveys are usually comprised of at least two separate parts: An appraisal of the exterior of the structure and drainage system, etc., and an appraisal of the interior of the building. Which comes first is usually dictated by the weather, with a hope that it might improve to allow an umbrella-less appraisal of the exterior façades, the grounds and the drainage-system. Whether outside, then inside, or inside, then outside, the important thing – apart from a detailed appraisal each time – is to look for supporting evidence *internally* of any anomalies detected *externally* or vice versa. For example, heads of doorways and window-sills, thought to be visually out-of-level externally (via historic settlement), can be more easily appraised by a '*closer-eye*' internally. And a sagging roof-ridge, seen externally, prompts a roof-void (loft) examination to check on the adequacy (and the sufficiency) of the purlin struts, etc.

Settlement cracks: This term refers to cracks in structural walls that are attributed to having been caused by a degree of ground/foundation movement, which either happened soon after the building was built, when the newly built structure was settling (shrugging) down, or is believed to have happened to an older building in recent times. The former is usually referred to as '*historic* settlement' and is less worrying than cracks that occur in older buildings that should not be settling down in their maturity. In these cases, depending on the actual appearance of the crack(s) and their position in the structure, it is wise to question whether the damage might be due to *subsidenc*e caused by nearby tree roots, or by a damaged, leaking drainage system close to the building's foundations. For further information relevant to this, see **Cracked-wall monitoring via Crack Monitors**, written herein under that heading.

Shared fences: See **Boundaries**

Sharp sand: This is an important ingredient of *concrete* and of sand-and-cement *mortar* mixes for chimney-top flaunching and floor screeds, etc., where greater strength and longevity are required. It can be visually differentiated from soft sand by its slightly larger, coarser, angular-shaped structure – and, of course, it *feels sharper* than *soft*, round-shaped sand.

Sheathing: See **Cladding**

Sherardizing: A manufactured protective-coat against corrosion, added to some nails and screws, by heating them in a revolving container with zinc dust at very high temperatures. Unlike galvanizing, the penetrated coating of zinc does not peel off.

Shingle: Small, rounded beach-pebbles, sometimes used to cover areas of a driveway.

Shingles: See **Wood shingles**

Shiplap boards: See **Weatherboarding**

Short grain: *Figure 65*:   This term refers to the visual, natural waviness of timber fibres (grain), as it may appear in relation to any section of straight timber; but when the

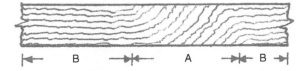

Figure 65 A = short grain; B = Long grain

(natural) waviness shows itself at one point to be obliquely across the timber's width or thickness, this point of waviness is referred to as *short grain* – which is considered to be a structural weakness that may fracture if an affected piece of timber is ill-placed.

Shuttering: See **Formwork**

Sill/Cill: See *Figure 35* on page 84: Either of these spellings are used in Building to refer to the horizontal base component of a window frame or doorframe. They may be of hardwood, softwood or uPVC., etc., and have an 8° to 10° weathered slope and an anti-capillary grooved drip-edge to their front undersides. Exceptionally, uPVC sills have a drip-edge in the form of a downturned front-edge – the ends of which are fitted with plastic L-shaped caps.

Silverfish: Small, silvery coloured, wingless insects (about 6mm/¼in long), of the *Lepisma* genus, found in interstices such as old book-bindings, behind starch-adhered wallpaper and in moist, damp places, such as between floor-coverings and substrate floors in bathrooms, etc.

Single-skin wall: A single-thickness (solid) wall, regardless of its thickness, which is sometimes referred to as a non-cavity wall.

Sink: Although this word has a general use, in plumbing and building terms, it also refers specifically to a *kitchen* sink – and not a **wash-hand basin**. However, regarding the latter, if too many syllables is a problem for people nowadays, it could be referred to as a *wash basin* or simply a *basin*. The main incongruities in amalgamating these terms are that a basin is normally much shallower than a sink and – more importantly – each has a different use.

Sinking fund: This is basically a managed savings-account that joint leasehold home-owners are obliged to contribute to every month, via a mandatory service charge. The idea is that the gradual build-up of this managed fund will prudently cover any unexpected building work that may be required – such as roof-repairs – apart from known maintenance items such as repainting communal areas, etc.

SIPS: These initials stand for *structural insulating panels*, which consist of dense-foam polystyrene sheets, pressurized/adhered between two weather-resistant OSB (oriented strand board) outer layers. Such panels, measuring in excess of regular sheet-material sizes of 1.2m wide x 2.4m long, apparently weigh between 50 and 80kg and offer a high structural and thermal performance, with U-values as low as 0.11 W/m2K. Their thickness is believed to be at least 100mm. They are being used as the inner, structural walls and roof-structures of single-storey dwellings and are made via CAD (computer-aided design) production. Their assembly-time on site (once the concrete-oversite is ready) is said to be less than a week. This inner shell is then ready to receive a site-built outer-leaf of brickwork or blockwork, etc., prior to being tiled or slated. Note that SIPS is another good example of **Modern Methods of Construction (MMC)**, listed herein.

Site: Any land or building, etc., occupied by builders, is usually referred to as a *building site*.

Site Agent: Nowadays, this usually refers to the on-site person in charge of a building contract, who is usually technically qualified to technician-level, more than being practically qualified to a craft-level. Ideally, such a position requires a mixture of both disciplines, but, understandably, this is difficult to achieve. Traditionally, Site Agents rose up from tradesmen (a term used before women entered the industry) with technical qualifications, to trades-foremen, on their way to becoming very experienced General Foremen. This hierarchical chain has now been broken and replaced by subcontractors and a Site Agent/Manager, often (sadly) with limited practical experience or knowledge of expected building-standards.

Site-cut pitched roofs: This self-explanatory term refers to the traditional method of roofing – still carried out to some extent on one-off, individually-designed properties – whereby first-fixing carpenters mark-out and cut the initial roofing components (mainly the common rafters) at ground-level, beside the building, before erecting the skeletal framework of the roof at scaffold-level. This allows them to work out other components (such as hip-rafters, valley rafters, purlin-bevels and lengths, etc.) more precisely, before cutting them – optionally on the roof-level scaffolding or down on the ground.

Skew-nailing: Nailing at an angle of about 30^0 to $45°$ to the nailed surface, through the sides of the timber, instead of squarely through the edge or face. Note that skew-nailing can also be done nowadays with a *framing-nailer* nail gun.

Skimming or setting coat: These terms relate to the fine finishing-plaster that is applied to rendered and/or plasterboard-lined stud-walls in a 3 to 5mm thickness, which is trowelled to a smooth finish.

Skirting boards: See *Figure 16* on page 39: This apt term for boards that *skirt* around the interior finished-wall and floor-junctions of rooms, was historically referred to as wainscot or wainscoting – although these archaic terms also applied to wall-panelling.

Essentially, skirting boards – like plinths – protect the base of walls and cover up the differential gaps and movement between the often-problematic junctures of different materials and constructions.

Skirting- or architrave-grounds: See *Figure 16* on page 39: Historically, up until the mid-20th century, timber *grounds* were fixed around the base of interior walls and the perimeter edges of door-linings, to provide a receptive fixing-material and (with regard to the skirting) a controlled guide for plaster-thickness and straightness of plastered surfaces. Regarding door-linings (which did not require a guide), they provided a defence against the initial moisture-absorption and distortion from the wet two/three-coat plastering. The timber grounds – which only seemed to be used on good-class work – were of a minimum 13mm/½ in. x 50mm/2 in. section of sawn or par (planed all round) softwood and had a 20^0 to 30^0 bevelled top-edge (that was meant to retain the plaster's edge, by sloping down to the wall).

Skylight sun-tunnels: These relatively-recent building products are circular tubes of various diameters from 250mm/10 in. up to about 500mm/19¾ in. They are also referred to as light tunnels or tubular skylights. This is because they need to be fitted into the roof (ideally in the south-facing side of a pitched roof), to allow the *light-tube* (yet another of their multiple names) to be accommodated in the loft area. Short-length, rigid light-tubes give a better light transmittance than the flexible tubes that often have to be used to avoid obstacles in the roof-void/loft area. They can also be fitted into flat roofs. The highly reflective tubes spread out the natural light received, into the room below. When I first witnessed the use of one of these minimal-diameter tubes (about a decade ago), in the ceiling of a windowless, dark bathroom of an empty bungalow being surveyed, I at first thought that a recessed, electric ceiling-light had been accidentally left on.

Skylight windows: See **Roof windows**

Slate-and-a-half slates: This description refers to a *single slate* which is 1½ times the width of the common plain slates used on a roof. Slate-and-a-halves are used alternately at the gable-end verges to create a staggered-joint bond. If the common slates are, for example, 200mm/8 in. wide x 400mm/16 in. long, the slate-and-a-halves would be 300mm/12 in. wide. x 400mm/16 in. long.

Slate battens: See **Tile** and **slate battens**

Slates: See **Roof tiles** and **slates**

Slate-tingles: Because slates are nailed centrally in their length (with two nails), when a broken slate needs to be replaced, it cannot be replaced and nailed as it was originally.

This is because the two nails holding the slate are covered up by the slate above. So, the technique used to carry out a repair, is to *rip* off the heads of the old nails (via a tool known as a *slate ripper*) holding the remaining piece of broken slate and fix a so-called **tingle**. This is a narrow strip of sheet lead, copper or zinc, of at least 25mm/1 in. width x about 150mm/6 in. length. It is fixed in a laterally central position on the slate-covered slating batten, and (after sliding the replacement slate up into position), the tingle's protruding-end is bent up onto the central, bottom edge of the replacement slate – as its only means of fixing support.

Note that a slated roof with a large number of tingles suggests that the roof may be suffering from *nail-sickness* – meaning that a large number of the slate-fixing nails have rusted away (just below their round heads) via metal-corrosion. Initially, such a conclusion will need to be confirmed by a close examination of the roof's underside, via the loft area.

Slatted shelves: As commonly used in airing cupboards, this refers to p/p (prepared/ planed) timber slats, usually of ex. 50mm/2 in. x 25mm/1 in. softwood, that are nailed or screwed to similar-sized timber bearers (fixed to the walls), with a slat-thickness space (gap) between each slat. Note that a slatted-shelf arrangement, around the three walls of a cupboard, would require the right-angled shelf-abutments to be splay-ended and splay-housed to each other. In such arrangements, these splay-housed joints are best glued with PVA adhesive.

Sleeper walls: This is a reference to low-rise, half-brick-thick, honeycombed walls built beneath timber wall-plates which support traditional, suspended ground-floor joists. Note that there would/should be a DPC bedded on the sleeper walls, beneath the wall plates, which would normally be at the same level as the DPCs bedded in the outer walls.

Slenderness ratio: This term refers to the height and/or the length of a wall in relation to the wall's thickness and whether any intermediate support is present (or required) in various forms. These intermediate supports may be in the form of brickwork or block-work cross-wall connections, piers or chimney breasts and buttresses. Such a subject enters into the realms of architecture and structural-design of buildings at the design stage, but Building surveyors have to develop an eye for violations to ratios of slender-ness, that may have been perpetrated by inexperienced builders or the plethora of DIYers around nowadays. Surveyors have to assess the structural integrity of a building under survey, within a relatively short space of time. An eye, therefore, has to be developed for cross-wall removals (especially where two rooms have been converted into one room); non-inclusion of point-load piers in half-brick-thick partition walls supporting a beam-end – and chimney-breast removal, etc.

Sliding-door lock: This traditional mortise lock has a concealed hook-shaped bolt, that (once the key is turned) emerges from the lock's face-edge plate, in a downwards locking action, to hook into the square-holed striking-plate that is housed in the door-lining- or frame-edge.

Slot-screwed expansion fixings and fittings: **1)** This refers to a method of fixing table-tops and/or worktop-counters from the underside, to counteract any likelihood of causing damage to such tops via thermal movement of the board- or sheet-material. Historically, on table tops, small, wooden *tongued-end buttons* were screwed at reasonable spacings on the underside – and their tongued ends were located into short-length and shallow-depth grooves in the rails. **2)** This principle is still used, but instead of tongued-buttons and holding-grooves, various-sized metal screw-holed plates or right-angled brackets are available, each with elongated screw-slots to allow for expansion or contraction to take place.

Snagging list: See **Maintenance period**

Snapped headers: This is usually a reference to stretcher bricks that were traditionally cut in half across their length (usually with a bolster chisel) and used as headers in a half-brick-thick outer skin of a cavity-wall to give the false impression of a more-solid, one-brick-thick wall simulating a Flemish-bond pattern of alternate headers and stretchers, whilst also economizing on the total number of (more expensive) facing bricks used. Such work – not easily detected – might also reflect an ancient, original concern about the stability of cavity walls, when compared with traditional one-brick thick, or one-and-a-half-brick thick solid walls. And, although cavity walls have now survived the test of time, it must be said, in all honesty, that they are not as structurally solid as the 225mm/9 in. or 343mm/13½ in. walls.

Soakaways: A traditional soakaway is an underground drainage pit which is built to receive rainwater from a roof and – if required – from hard ground-surfaces, such as a driveway. The underground pipework serving the soakaway is laid to a 1 in 40 gradient from P-trapped gullies. The soakaway must be large enough not to be overwhelmed by rainwater, until an adequate dispersal has leached out/percolated into the surrounding ground. Such systems are to prevent storm/surface water from entering and overwhelming the main foul-water sewage system. Therefore, as described above, a soakaway may be either **1)** an unseen/undetectable cube-like hole in the ground, filled with broken bricks and stone, covered over with soil (and possibly vegetation); or, **2)** a sophisticated, but simple arrangement of linked-up modular, storage/soakaway cells. The perforated polypropylene, lookalike-crates can be built-up to suit the required pit-size. And they can either be used as a common soakaway or, if wrapped around with an impermeable geo-membrane, they can be used as an upgraded storage system for stormwater. Note that brick/stone-filled soakaways, as described above at **1)**, seriously reduce the soakaway's capacity.

Soakers: *Figure 66:* Pliable pieces of metal, usually of lead, copper or zinc, fashioned into right-angled shapes to fit under the outer edges of slates or plain tiles that abut any roof-projections, such as chimney stacks and party-wall projections. The soaker-abutments to the wall must be covered with separate, stepped lead-flashings, which have top-turned edges that are (or should be) lead-caulked into stepped, raked-out brick-joints, prior to being pointed.

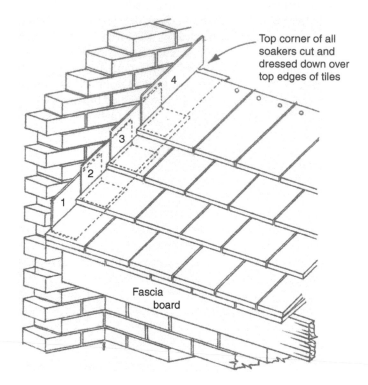

Top corner of all
soakers cut and
dressed down over
top edges of tiles

Fascia
board

Figure 66 An example of four lead 'soakers', showing how they should be overlapped against a wall-abutment. Note that the soakers must be covered with stepped, lead-flashing (as illustrated herein at Figure 30)

Soffit: The underside of a beam, ceiling, lintel, staircase, or an eaves' projection, etc.

Soffit boards: See *Figure 67* on page 231: The underside of eaves-projecting-rafters were originally lined with a lath-and-plaster soffit, which was superseded by timber-boarded soffits in the early 1900s; but, because of wet-rot deterioration, due to their vulnerability and poor painting-maintenance, they (and the fascia boards) started to be covered with uPVC in the late-20th century. Nowadays, soffit boards and fascia boards are usually entirely of uPVC, without any timber substrate.

Soft sand: This is recognizable by its small and soft (unsharp), spherical grains, which can be handled comfortably, as in the building of a sand castle. It is mostly used in the industry for making bricklaying-mortar and pointing. It should not be used for making concrete, which requires *sharp* sand.

Softwood: This term is a commercial description for the timber used in industry, that has been converted from needle-leaved, coniferous evergreen trees, belonging to a botanical group known as gymnosperms. Occasionally, the term *softwood* is contradictory to the actual density and weight of a particular species. For example, *Parana pine* is a soft-wood that is quite heavy and dense, like most hardwoods.

Soil and vent pipe (SVP):   This 110mm/4¼ in. Ø (diameter) plastic or cast-iron pipe is also referred to as a *soil stack*. It rises up *through* a building (although originally *outside* of a building) to receive soiled, foul-water from WCs., sinks and basins, etc. via branch pipes en route to the underground drainage system. It also provides a necessary vent to the drainage system. Note that – as per the Building Regulations – the open-top of the SVP should be at least 900mm/35½ in. above the uppermost part of any opening-window. However, this rule does not apply to opening-windows that are 3m/9ft 10 in., (or more) laterally from the top open-end of a soil and vent pipe.

Soil (or foul) drain: This usually refers to the underground drainage that carries sewage or industrial effluent to the main sewer, cesspit, septic tank, or sewage treatment plant. Such pipework, traditionally, was of SGW (salt-glazed ware or *earthenware*), but now-adays it is of uPVC. The diameter of these pipes is usually 110mm/4¼ in., or 150mm/6 in.

Soil pipe or **soil stack:**   See **Soil and vent pipe (SVP)**

Solar reflecting roof-surfaces: This refers to flat-roofs that are covered (weathered) with dark-coloured materials such as asphalt or black, bituminized 3-layered roofing-felt. The addition of special surface-finishes such as white spar (rock) chippings, or heat-reflecting paint, can considerably reduce the surface-heat during sunlight.

Solar voltaic panels: These roof-panels, facing in a southerly direction, if possible, create electricity by absorbing sunlight via their photovoltaic cells, which generate *direct current* (DC) *energy* and convert it to safer *alternating current* (AC) energy. This energy source then flows through to the dwelling's mains electrical distribution panel.

Soldier arches: As the name suggests, such arches are comprised of bricks *stood upright* on their face-ends, *side-by-side in a row*, with their longer and wider faces bedded together with vertical mortar-joints. Historically, because this type of arch lacked a geometric curve or camber – and was therefore only partly self-supporting by its overall compact-width within the mortared brickwork side-abutments – two small-diameter ferrous-steel reinforcing bars, with U-shaped, eyelet-ended steel-stirrups hanging from them (like curtain-hooks), were bedded in the mortar-bed above the arch. As the work proceeded, the stirrups were slid along the bars, to be mortar-bedded between the soldiers. Nowadays, such arches are usually supported by being built onto the outer flanges of *galvanized-steel* lintels, and the historic, erroneous practice of using *ferrous-steel* reinforcement has ceased.

However, when surveying old-build properties that display soldier arches that may contain built-in ferrous-steel reinforcing, one should look out for mortar-joint cracks between the soldier-bricks and/or the mortar-bed joint immediately above the soldier arch, because, if the U-shaped stirrups and the bed-joint reinforcing bars have corroded and expanded in size via progressive rust-expansion, this may well have built up enough to cause structurally-damaging, lateral movement of the brickwork. One case with this phenomenon, witnessed a few years ago, was so severe that the degree of lateral movement in the soldier-arched, windowed outer-leaf of a cavity-wall had caused the left-hand *quoin* (the outer corner of the wall) to bulge noticeably in its mid-area height by approx. 16mm/⅝ in.

Sole or **Sill plate:**   These archaic terms refer to a *floor* plate, as commonly used in timber stud-partitioning.

Solid-core doors: These are usually faced with plywood and are also referred to as *flush doors* – although flush doors may not necessarily have a solid timber-core. Those doors that do, are usually rated as FR doors (fire-resisting doors). FR doors carry a stability/integrity rating which is expressed in minutes, such as 30/30, or 60/60. *Stability* refers to the point of collapse, when it becomes ineffective as a barrier to fire spread. *Integrity* refers to holes or gaps concealed in the construction when cold, or to cracks and fissures that develop under test. Nowadays, the doorframe rebates (opposite the door edges) might be required to be shallow-grooved to receive one or two (side-by-side) intumescent strips.

Solid doorframe: See *Figure 57* on page 166: This terminology distinguishes a doorframe with *stouter* timber sections and rebated recesses to receive the door-edges, as opposed to *thinner* timber sections with *planted* (separate) doorstops. However, the reference to being *solid* is not normally used. It is just referred to as a ***doorframe***; and the thinner version with planted doorstops is referred to as a ***door-lining***.

Solid floors: This description is used for floors that are not constructed of timber joists, but are of concrete in its various forms. They may be at ground-floor level, referred to as the *concrete-oversite* or *ground-slab* (usually involving a DPM – *damp-proof membrane*), or they may be suspended upper-floors, of reinforced concrete, with various other materials laid upon them.

Solid strutting: This is sometimes used on suspended timber-joisted floors to give additional strength via mid-area connection between joists. Strutting removes the individuality of each joist and creates an equal distribution of the floor's load, thus preventing the joists from bending sideways. Struts should be used where floor-spans exceed 50 times the joist thickness. Therefore, with 50mm/2 in. thick joists, a single row of central struts should be used when the span exceeds 2.5m/8ft 2½ in., and two rows are required for spans in excess of 5m/16ft 5 in. and up to 7.5m/24ft 7¼ in.

The present-day practice of floor-strutting with solid timber struts was frowned upon traditionally as adding unnecessary weight and creating an inflexible floor – but nowadays, there are at least two exceptions where it is recommended: **1)** Section 5 of the *New Build Policy's Technical Manuals* recommends that solid strutting should be used instead of herringbone strutting where the distance between joists is greater than three times the depth of the joists; and **2)** the **TRADA** (Timber Research and Development Association) document: '*Span tables for solid timber members in floors, ceilings and roofs for dwellings*' recommends that solid strutting be used for certain long-span domestic floors.

Note that the *New Build Policy's Technical Manuals* mentioned above are registered-builders' guidance notes supplied to them by Zurich Municipal Insurance (ZMI), whose surveyors monitor their building work during construction. Upon completion, ZMI issue a certificate guaranteeing the building's fitness for a period of ten years.

Solid walls: This term refers to **single-skin** (single-leaf), non-cavity, brick-, block-, concrete- or stone-walls, regardless of their thickness. It does not necessarily refer to their condition or strength.

Solvent-welded joints: This describes the modern-day joints that are made between the male- and female-ends of plastic pipes. The interconnecting surfaces of each pipe are thickly coated/brushed with a solvent before being coupled up and held for a few moments.

Soot-ash damage: See **Deliquescent soot-ash staining to chimney-breast plaster**

Sound proofing: See **Acoustic construction**

Spall/spalling: 1) Brickwork: Certain soft bricks, such as *red Rubbers*, made from sandy clay, gained their name from being easy to cut and *rub* into voussoirs (tapered arch-brick shapes), but they also have a disadvantage by sometimes being too soft and absorbent, which makes them prone to surface-damage from *spalling* (surface-splintering). This damage can (and does) repeatedly occur when the elevational faces of such brick walls are wet from driven rain and are then subjected to frost. The absorbed moisture freezes in random areas and expands upon changing to ice, causing tiny substrate cracks. Then, once the partly-frozen surfaces start to thaw, the cracks give way to sand-like disintegration and small, tablespoon-shaped craters are left in evidence of the occurrence. Over time, these craters are usually susceptible to further attack and instances of half-brick-depth damage are not uncommon on some buildings.

2) Plaster: The term *spall* is also used when a small area of finishing plaster breaks away (*surface-splinters*), or partly breaks away, from a plaster-boarded ceiling (or a timber-studded wall) which has been fixed with galvanized or sherardized nails – and the nails have been overdriven and have broken through the board's outer paper-surface. This damage, via inevitable future vibration, causes a small, roughly circular patch of plaster to break away (or partly break away and hang loose) from the nail-head area.

Note that such occurrences are less likely to happen nowadays, since nail fixings were superseded by screw-guns using bugle-head shaped screws.

Spandrel: This is a reference to the triangular-like area/space above each side of an arch, which needs to be filled with carefully-cut masonry before complete lines of masonry can be laid above it.

Specification: See **National Building Specification**

Specification for Small Works: This title refers to a simply-worded, unambiguous, but detailed description of relatively small building jobs/works to be carried out on a property. Although brevity should be the aim, one must convey clearly what is required to be done and, if possible and relevant, what materials, etc., should be used. The idea being that such Small-Works' specifications can be given to your builder – or to an unknown builder for them to supply you with a *quote* for the stipulated work – as opposed to them giving you a ball-park *estimate*, that may be inflated upon completion of the work.

Spindles: See **Balusters**

Spiral stair: A stair/staircase that describes a helix around a central column.

Splashback/Backsplash: The first term has been used in the UK for many decades, meaning an impervious material such as ceramic tiles or glazed panels, etc., fitted around the perimeter walls immediately above the worktops of fitted kitchen-units, to protect such walls from splashes. The second term (backsplash) seems to be another Americanism.

Splay-topped copings: See *Figure 14* on page 30: This refers to so-called *coping stones,* made of natural or artificial stone, or precast concrete, that have a shallow-angled *weathering-splay* of about 15° on their topsides. They may be flush to the brickwork in their width, or have small projections that accommodate grooved, drip-edges on their underside.

Spliced timber: **1)** This usually refers to small areas of damaged or wet-rot timber that has to be cut away to receive a dovetail-shaped splicing of sound timber; **2)** This term can also apply to *spliced jointing,* such as is used for lengthening *roof purlins* or *ridge boards,* etc.

Spot board: This refers to a square *mortar board,* with sides of about 760mm/30 in. long, usually made of 18mm/¾ in. thick shuttering plywood – and used by bricklayers – or a larger square board sized 900mm/3 ft. – and used by plasterers.

Sprigs: See **Glazing sprigs**

Spring/Sprung: Warping that can occur in timber after conversion and seasoning, producing sprung, cambered or segmental-shaped edges adjacent to the wide face of the timber. In practice, joists and rafters, so affected, should always be placed with their sprung edges uppermost, thereby allowing the weight of the floor- or roof-loads to stabilize the misshapen edges – and possibly straighten them out to a certain degree. In traditional site-cut and pitched roofing, the sprung top-edges of any common rafters were pushed or cramped down and skew-nailed to the mid-area purlins.

Springing line: A horizontal reference line at the base of an arch (where the arch *springs* from), that relates to the initial setting-out, which may be simple, as with a *segmental* arch, or complex, as with a *semi-elliptical* arch with three or five geometric setting-out centres.

Sprocket-formed eaves: *Figure 67*:   This terminology refers to 600mm/2 ft. long wedge-shaped pieces of ex-rafter material (sprockets), fixed on top of each rafter at the eaves to create a bell-shape appearance (an upward tilt) to the roof slope. This effect is also achieved – but with a greater visual display – by fixing short offcuts of rafter-material – set at a shallower angle – to the sides of each rafter-base, just above the wall plates. Apart from aesthetic reasons, sprockets might be added to reduce a steep roof slope in order to ease the flow of rainwater into the guttering.

Spur: An electrical socket-outlet that has been added (like a short protruding branch from a tree) to a *ring main* outlet. Note that too many spurs on a ring main might contravene Part P of the Building Regulations.

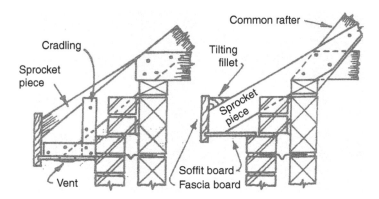

Figure 67 Sprocket-formed eaves

Squaring methods used on site with laser levels: Optical instruments with integral laser-plummet facilities that emit a vertical beam from their tripod position, down onto a corner setting-out peg, can be used to establish 90^0 right angles required for setting out walls at foundation level, or, less expensive laser squares can be used. Also, when the longest side of a right-angle in any setting-out does not exceed about 20m/65ft., the angle can be set out accurately enough by using a simple method of geometry, known as the 3:4:5 method, or alternatively, the 5:12:13 method. These methods are explained herein under the heading: **Right-angle forming methods.**

Squint quoin: An uncommon corner of a building that is not a right angle and is therefore less or more than 90^0. Such quoins can be formed with special-made bricks (with one splayed end), or the square-ends of the facing bricks are allowed to protrude or recede on each course – with the resultant feature of small, alternating brick-triangles that project on one side and the other, or similar, small triangles that recede. Purpose-made bricks are best, of course, but are much more expensive.

Stable door: **1)** An exterior-type, ledge-braced-and-boarded (L&B) door to a stable, that is comprised of two halves that meet and/or separate on a rebated-edge running horizontally through the middle rail. Each half-door, which hangs on side-hinges, one-above-the-other, is separately bolted and staple-locked; **2)** A door, very similar to the above style, but which has been adopted for external-use in dwelling houses. Such doors, in their top halves, are usually multi-glazed (via an arrangement of glazing bars) and have a protruding *drip mould* fitted to their bottom edges.

Stack: In building terminology, this can refer to a chimney stack, or vertical pipework.

Staff-, Stop-, or **guard-beads:**   These terms refer to the interior beaded-edges of traditional boxframe windows – and such windows are still very much in vogue. These beads retain the inner sliding-sash, but they are removable for renewing broken sash cords. In coastal areas, the guard-bead fixed to the sill is usually deeper and should be tongued into the sill, acting as a weather-bead. It might also contain an anti-capillary feature. Unless deep guard-beads exist, wind-driven rain can be forced up-and-over normal-sized stop-beads. Note that the term *bead* refers technically to a semi-circular shape, with a one-sided quirk.

Staffordshire blues: These engineering bricks are highly regarded in the UK for their high crushing-strength and durability. As the name suggests, they are a deep blue colour.

Staggered joints: From a structural (and sometimes) a visual point of view, end- or side-jointing of boards or sheet material, etc., should be staggered. To make the point, a brick wall would not be very strong if all the perp (perpendicular) joints were one above the other.

Staggered-stud partitions: Such partitions, used for reducing the transmission of sound between rooms, have been around for many years now. They are also referred to as being of *discontinuous construction.* The reason for this nomenclature is because the vertical timber *studs* (posts) are staggered on each side of wider-than-usual floor-plates and head-plates – by being alternately flush to the face-edge on one side and then to the face-edge on the other side, whilst also being positioned in-between each other on each side. This simple technique (simpler to do than to describe) creates a wavy central space between each staggered row of studs, which allows a sound-reducing quilt-insulation screen to be internally housed.

Stair or stairs: In reference to a flight of steps, *rising* (or *falling*) from one level to another, these terms have a singular- or plural-form, seemingly emanating from historic misuse between Scotland and England. However, nowadays, south of the border, flights of steps are more commonly referred to as *stairs* or a *staircase* (singular), or *flights of stairs* or *staircases* (plural). Although, historically, the term staircase meant the space within which a stair was fixed – but this space is now referred to as a *stair well*, or a *stair-well* – a term which originally referred to the potentially dangerous open-*well* created in the mid-area of a three-quarter-turn staircase. However, to add to this confusion, the latest amended AD K1 Stair Regulations refers to *stair* as being *a succession of steps and landings.*

Staircase: See **Stair** or **stairs**

Staircase-undercarriages: These are used on geometrical staircases that have an open, cut, mitred and bracketed outer string-board, as opposed to a common, closed outer string-board. Such strings, by virtue of the triangular-loss on their top edges, usually require rafter-like sloping timbers (undercarriages) on their underside, to compensate for their structural loss on their top, compressive edges. The undercarriage timbers, three in all, were used to carry the stair's soffit (ceiling). Also, a thin bead-edged lining board was usually attached to the face-side of the outer-undercarriage, to conceal its rough, sawn state, prior to soffit-plastering.

Staircase-width: Contrary to original building regulations – which gave 800mm/31½ in. as the minimum unobstructed width for the main stair-flight in a private dwelling – no recommendations for minimum stair-widths are now given. However, stair-designers must bear in mind the requirements for stairs which: **1)** form part of a means for escape, as per *Approved Document B: Fire safety*; and **2)** provide access for disabled people, as per *Approved Document M.*

Stair-design formula: As per the Building Regulations' Approved Document K1 (AD K1), reference is made to three categories of stairs (given here separately under **Categories of stairs**), all of which must be designed to meet the given formula of *2R + G* (twice the rise + the going), which must equal not less than 550mm, nor more than 700mm.

Stairwell: Traditionally, this term referred to the open, central space created by quarter-turn, half-turn or geometric stairs, that turned around two or three walls, thus creating a central, *open-well*, but, although this is still technically true, the term stairwell is used nowadays in reference to the whole stairwell space *before* the second-fixing operation of fixing the staircases takes place.

Stanchion: A vertical, steel column, usually H-shaped in cross-section, that supports a superimposed structure, usually in the form of RSJs (rolled-steel joists) as part of a floor.

Star dowel(s):   Cross-headed and cross-pointed alloy-metal dowels, used traditionally to secure the glued comb-joints of BWMA-type casement sashes/windows.

Steel herringbone-struts: At least two types of galvanized-steel herringbone struts are produced as an alternative to traditional wooden strutting. The first type, by Catnic-Holstran, have up-turned and down-turned fixing lugs, used for fixing to the sides of the timber joists with minimum-length 38mm/1½ in. round-head wire nails.

The second type, with the 'BAT' trademark, has two-eared, pointed ends which simply bed themselves into the sides of the joists when pushed in at the bottom, by being pulled down firmly at the top. This time – as opposed to fixing wooden strutting – fixing is done (more safely) from the floor below. One disadvantage with steel strutting, related to them being manufactured to suit joist-centres of 400, 450 and 600mm/16, 18 and 24 in., is that there are always one or two places in most floors that do not conform to size and require reduced-size struts. When this occurs – which is more often than not – it is necessary to use a few wooden struts in these areas.

Prior to fixing the struts, the joists running along the sides of each opposite wall, will need to be packed – with wedges, traditionally, but timber-shims can be used – and nailed (or adhered) directly behind the line of intended struts. As stated under **Herringbone strutting,** herein, fixing of these is usually done from above.

Steel-reinforced soldier-arches: See **Soldier arches**

Stepped cavity-trays: See *Figure 18* on page 46: When a pitched roof's gable-end abuts a flank wall of cavity-wall construction, the abutment of the two-opposing roof-slopes must be weathered by stepped, lead-flashings (or other impervious material), dressed over pre-fitted lead-soakers (as described under the **stepped-flashing** heading). But more than that, because the abutted wall is a cavity wall, the cavity must be fitted with purpose-made, stepped cavity-trays, that are seated immediately above the stepped lead-flashings. This provides the abutted wall with a DPC (damp-proof course) – to inhibit descending damp.

Stepped-flashing: See *Figure 30* on page 69: These weather-protection covers can be of copper, zinc or bituminous felt, but are more commonly formed from sheet lead. They are typically (but not exclusively) used at the angled faces of the side-brickwork of a chimney stack, where it emerges through a pitched-and-clad roof. When slates or plain tiles have been laid – and lead *soakers* have been placed between them, right-angled up against the brickwork (as shown in *Figure 80* on page 269) – the soakers are then covered and protected by a separate, stepped flashing – the top-edges of which are turned and dressed into pre-raked-out mortar joints. Folded lead wedges are driven in intermittently, ready for the bricklayer (or roofer) to point with sand-and-cement mortar.

Sterling board: See **Oriented Strand Board (OSB)**

Sticks or stickers: These terms refer to equal-thickness timber battens (sticks) that should be laid between stacks of timber or sheet (board) material, that are being stored or awaiting use.

Stock bricks: This historic nomenclature refers to bricks that were mostly available to a particular district. So, the well-known *London stock brick*, a yellow/brown/black-mottled brick, made from Kentish clay, was much used in London until faced Fletton bricks arrived.

Stool/stooling: See *Figure 3* on page 6: This is a reference to the flat, top-ends of stone or precast concrete, *weathered* (sloping) window sills, that are left as small, *seating* areas, upon which the brick-reveals of a window-opening are bedded. The small, transitional sill-sides of stooling are usually cove-shaped to the weathered slope, as illustrated. Sill-end stoolings give a neater appearance to the coursed brickwork bedded upon them.

Stopcock/screw-down valve: This is fitted in a waterpipe-run which is under high pressure. Such fittings are available to suit copper, plastic, lead and low-carbon steel pipes. Note that it is important to remember that a stopcock must be fitted only one way round, with its stamped arrow pointing in the direction of the flow of water. The mechanics of this direction allows the washer-fitted metal-jumper to be forced up from its seating-seal, when the spindled tap is turned to release or continue the flow.

Storey-heights: Part A1/2 of The Building Regulations gives storey heights, but defines them as being controlled by the provision of effective lateral support to the walls, in the form of being tied to the walls, or of being suspended floor-bearings on the walls.

Thereby, if the ground floor of a building is not deemed to provide effective lateral- support to a wall, the *storey height* should be measured from the base of a wall. Otherwise, the ground-floor storey height should be a maximum of 2.7m/8 ft.10⅜ in. from the floor-top to the underside of the ceiling. And the first- and second-floor storey-heights are given as being less than this, by virtue of the same measurement being measured from the *underside* of the floor to the underside of the ceiling. So, with these regulatory requirements in mind, designers and architects using dry-lining techniques on the walls could use the available standard lengths of British Gypsum's Gyproc Wallboard (positioned to achieve vertical joints), measuring 2.7m/8 ft.10⅜ in. x 1.2m/3 ft.11¼ in., of 12.5mm/½ in. thickness, with tapered- or square-edges.

Storey-numbering (in the UK and the USA):   Apparently, a difference in numbering exists between these two countries, but only when there is a *basement* or *cellar* involved. In the UK, *ground-floor level* is of the same nomenclature, regardless of whether there is a subterranean (or partly-subterranean) *basement* or *cellar*. But in the USA, what the UK regards as the ground-floor level, becomes the *first-floor level* when there is a sub-terranean (or partly-subterranean) basement or cellar. So, in the UK, the first-floor level will always be above the non-subterranean ground-floor level, but in the USA, it could be one more floor above; i.e., the equivalent of the UK's second-floor level.

Storey rod and Gauge rod: (**1**) A straight, softwood batten, usually of 50mm/2 in. or 75mm/3 in. x 25mm/1 in. section, used by bricklayers, carpenters, or others on site for checking or establishing precise storey heights between floor levels (for stair-making, or concrete stair-building); or **2**) a similar batten used as a gauge rod with 75mm/3 in. incremental marks or sawcuts across its face for checking the incremental build-up of the brick quoins (return-corners) of a building as they are being built. Such gauge rods require short-length datum battens to be fixed to the lower levels of the brickwork at each corner, upon which they can be rested each time a bricklayer wants to check the incremental rise of his work. Failure to use some form of *gauge-rod technique* can have disastrous consequences on a building's floor-levels, etc.

Stormproof hinges: See **Cranked hinges**

Straight brick-arch (with tapered-voussoirs):   Such arches are still quite commonly used as features over windows within face-brickwork façades. Similar to the simple geom-etry for segmental arches, the span on the springing line is bisected to provide a vertical centre line. The centre point below this line is chosen (or dictated by an elevational drawing), but is only needed as a radiating point for the tapered voussoirs – and is not needed to describe arched curves – because the arch has no curved top and (seemingly) no curved *bottom* (soffit). However, a completely straight-bottomed arch (whether of fanned voussoirs or soldier-courses) can appear to be sagging – and was (should be) built on a wooden *turning piece* with a cambered top-surface and a minimal, seg-mental rise of 3mm/⅛ in. for every 300mm/12 in. of the span-width, i.e., 900mm/3 ft. span = 9mm/⅜ in. rise. Traditionally, such a minimal segmental curve was produced on a solid piece of wood called a *turning piece* – made by a site carpenter.

Straight flight: A staircase without winding, tapered steps at the top or bottom.

Straining piece(s): See *Figures 11 and/or 45* on pages 22 and 116: This refers to: **1)** horizontal timber-plates (with a sectional size of at least 100mm/4 in. x 63mm/2½ in.), whose square-ends are fixed up-against the ends of angled struts in a roof void. The straining pieces are nailed to the top of the ceiling-joists, to take the strain of the roof-purlin's load when the purlin-struts are angled down to each of their plumb-cut ends; or **2)** similar horizontal plates, nailed to the top- and bottom-surfaces of a horizontal shore (such being a component of a **flying shore**), to take the strain of pressures transferred to them via the upper and lower 45⁰ angled struts.

S-traps and **P-traps:** These critical plumbing items, attached to the various diameters of pipes serving WCs, wash-hand basins and sinks, etc., are so configured to retain (trap) a certain amount of flushing water above the inverted top of the U-bend, thereby preventing the likelihood of foul air/objectionable smells rising up from the so-called *foul drain*.

Stretcher-bond brickwork: *Figure 68:* An elevational pattern of bricks comprised of repetitive rows (courses) of *stretcher* bricks which are laid centrally over every perpendicular (*perp* or *perpend*) joint below. This bond is created at the quoins by the necessity of alternating header bricks that form the return-end of the wall or the beginning of another right-angled wall. Note that if wall-lengths are not designed by the architect/designer to *co-ordinated* brick-sizes (that include the mortar-joints), it may be necessary for the bricklayer to include ¾-length bricks at some strategic point in a wall. See references to *co-ordinated brickwork* under the **Brick-sizes** heading.

Stretcher brick: In brick-bonding terminology, this is a reference to the length of a brick.

Striking plate: A thin metal, T-shaped plate, with a central, rectangular hole to receive a door-latch – or with two holes in the case of a *door-lock* – with countersunk screw-holes on each side.

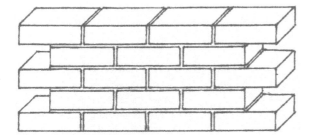

Figure 68 Stretcher-bond brickwork

String: In carpentry and joinery, this refers to the board or boards that house and support the end-grain sections of the treads and riser-boards of a staircase.

String course: This refers to a decorative or ornamental horizontal-band of brick- or stone-work on the façades of walls, that can be on a front elevation, or run all around the building. They are about three or four brick-courses high and usually coincide with the concealed position of the upper floor-level or levels. In its simplest form, string courses appear as a slightly recessed band of brickwork, but they can also be seen as brick soldier-courses, etc.

Strip foundations and **Trench-fill foundations:**   See *Figure 22* on page 56: **1)** Strip foundations is the technical term for traditional, concrete foundations, that were laid with a *width* equal to not-less-than three-times the thickness of the wall to be built upon them – but this guide-rule (written in Building Bye-laws prior to their consolidation into the Building Regulations in 1985) stated that this must not make them less than 225mm/ 9 in. deep for load-bearing walls. And the concrete's thickness was also required to be not less than the wall thickness and not less than 225mm/ 9 in. depth. Up until about the mid-20th century, foundations were comprised of unreinforced concrete, but then at least two 13mm/½ in. (or two 18mm/¾ in.) diameter steel-reinforcing rods started to be specified for positioning in the concrete's tensile area (near the bottom). The critical question of a foundation's depth below ground has always been subject to the nature and quality of the substrate, but – because of frost damage – it became established that their depth should not be less than 915mm/36 in.

   2) *Trench-fill/strip foundations* refer to a transformative idea in the late twentieth century, that reasoned it to be quicker and cheaper to fill (or *almost* fill) foundation-trenches with concrete, rather than awkwardly lay strip foundations at the bottom, for bricklayers to work awkwardly (at a slower pace) within the confines of a trench. Of course, the other ingredient that made the concept of trench-fill foundations possible was the availability of ready-mixed concrete, which has been transportable to building sites for many decades – apart from being produced in bulk on large building sites. Finally, the depth of the trench-fill foundations has increased from the minimum depth of 915mm/36 in. given above for older strip foundations, to now being not less than 1m/ 39⅜ in.

Strip hinge: This term usually refers to a piano-type hinge in one continuous length.

Struck joint: See *Figure 59* on page 191: This traditional mortar-finish to brickwork joints that had been raked-out to a 12mm/½ in. depth, for *pointing* (applying a neater, stronger mortar-mix) was applied with a small, pointing trowel and neatly bevelled slightly inwards at the top, and usually with a slight projection at the bottom, that was trimmed off finally with a small wooden straightedge and a so-called *Frenchman knife*. This procedure followed the pointing of the perpendicular (perp) joints. See other forms of pointing covered here under **Pointing**.

Structural appraisal of low-rise buildings: All building surveyors (and others interested in the condition of a building) must *develop an eye* for level, plumb and slightly bulging/ slightly hollow surfaces when appraising a building's condition. This should be a separate operation, involving a different mindset to the detailed observation of windows, doorways, guttering and downpipes, etc. on the exterior examination. If necessary, binoculars should be used, but the wider, open view obtained without them can be more revealing – less inhibiting.

Study the length of the roof's ridgeline; is it straight, sagging or bulging sideways? If it is sagging, it might suggest roof-spread, so study the eaves' projections on each side. If the fascia boards appear to be bowing out (when viewed from end-to-end), or the soffit boards have a suspicious mid-area bowed-gap against the walls, this is usually a sign of roof spread that will need to be investigated in the roof void during the interior inspection. If the ridge line appears to be bulging, this is usually a common phenomenon, likely caused by the ridge board at that point being supported by an underlying party/ firebreak wall being built-up to the roof's underside. Finally – although uncommon – one must realize that roof-spread can push the upper regions of the load-bearing walls over to out-of-plumb angles of lean.

Next, the elevations should be eyed for plumbness, especially if appraising an older-built property. And any suspected angles-of-lean or bulges should be further checked and recorded. In this respect, a spirit level and a transportable straightedge should be used.

Finally, regarding historic settlement (often referred to as 'shrugging down') that can be seen to have affected a large number of older buildings, this can usually be detected by appraising the heads and the sills of windows and doorway openings to the exterior façades.

Structural Engineer: A person qualified in structural mechanics and building-design, who usually designs and/or approves the structurally-calculated elements of an architect's initial inspirational work. Such professionals have invariably also developed a good analytical mind in the understanding of – and the remedial action required for – structural problems and defects.

Structural Engineer's Survey: Structural Engineers may be asked directly by a client to carry out a survey on a property, especially when alarming cracks, bulges or partial collapse of walls, etc. occurs. Or they may be recommended to the client by a Building Surveyor, when the latter knows that the matter is serious, but wants it to be endorsed by a specialist. However, if either surveyor's findings are related to subsidence, both will know that the client's Buildings' Insurance Company will want to have this confirmed by their own appointed surveyor.

Structurally-graded timber: Such timber is now commonly specified for structural uses and is covered by BS 4978:   'Timber grades for structural use'. Certain standards and criteria are laid down regarding the size of knots, slope of the grain, etc., and the way in which the timber is to be examined and stamped accordingly in designated grades.

Structural Survey: See **Building Inspections**

Structural walls: This term can apply to the interior walls of a building, as well as the exterior walls. So, regardless of the thickness of interior walls and whether they are built with brickwork, blockwork, timber- or metal-studs – they must only be determined as load-bearing walls by whether they carry an *additional* load apart from their own weight. Note that such additional loads require careful checking on the floors (or in the roof-voids) above.

Strutting: See **Herringbone strutting** or **Solid strutting**

Stub tenon: In joinery, a tenon that is cut short and is positioned in a blind mortise, i.e., it does not pass through, usually for aesthetic reasons of not wanting to show the end-grain.

Stuck or planted mouldings: See *Figure 12* on page 26: Traditionally, if a moulded edge was referred to as being a *stuck* moulding, it meant that the moulding had been *formed* on the edge of the timber – either by a router or a spindle moulder. But if it was referred to as being a *planted* moulding, it meant that it was a separate moulding, to be fixed by nails or panel pins.

Stud partitions: See **Metal-stud partitions** or **Timber-stud partitions**

Sub-contract: Such contracts are controlled by the main contractor of a building contract, when certain specialist tasks – such as installing a sprinkler system; supplying and fixing metal balustrading, etc. – have to be carried out.

Subfloor: As the term suggests, this is a reference to a *lower* floor, in relation to an upper, *finished* floor. One may be of sand-and-cement screed, the other of parquet flooring.

Sublet: Main contractors are usually allowed to sublet parts of the work to sub-contractors, but sometimes this has to be approved by the client or the architect.

Subsidence: **1)** This refers to a usually slow sinking or erosion of the ground (or one part of the ground) below the foundations of a property. This is often detected by stepped, horizontal cracks in the mortar-joints of the outer (and/or inner) brick-walls, when the stepped cracks are stepping down from a corner (quoin) of a wall – and the stepped, bed-joint cracks are wider than the stepped, vertical-joint cracks. **2)** Subsidence may also be seen in the form of a roughly-vertical crack in the middle (or near middle) area of a wall, where the crack is wider at the bottom than at the top and emanates from ground-level or paving-level, i.e., without being interrupted by the DPC. The fact that a vertical crack might be detected *below* and *above* the damp-proof course, is usually a strong indication that a degree of subsidence has taken place. And if such cracks are wide at their base and narrow or hairline at their top and are roughly in the middle area of, say, a flank wall, rising up to a considerable height, it seems very likely that mid-area subsidence has occurred that has, in building terms: 'broken the wall's back'. Note that (as mentioned above), if such a crack had not been seen below the damp-proof course, it would be more likely to have been caused by thermal expansion and contraction of the flank wall above the DPC.

Subsoil erosion and subsidence damage: Another form of foundation-damage that can trigger subsidence is subsoil erosion. This refers to the gradual breakdown/washing away of the subsoil of a foundation via a nearby cracked/leaking drain, a leaking underground water-pipe or a damaged, leaking (or blocked and overflowing) rainwater gully, up against the subsidence wall. Note also that nearby, foraging tree-roots often cause a lot of damage to drains, causing them to leak. Roots, in their initial, miniscule thread-like state (and acting intuitively, like mycelium, seeking moisture), are often discovered to have crept through cracks in a manhole's eroded mortar joints and to have caused damage (or blockage) to aged earthenware drainage systems and traditional open-channel entry/exit points in brick-built inspection chambers.

Substrate: In building, this is usually a reference to the structure or material below the finished surface.

Subterranean rooms: See **Basement rooms and/or cellars**

Suffolk latch: See **Thumb latch**

Sulphate attack: See **Leaning chimney-stack phenomena**

Sump-pump: A small, submersible pump, used below floor-level in an equally-small, square floor-sump. These are activated as and when problems with damp-penetration are occurring in subterranean rooms or wet cellars, via a high water-table in the surrounding terrain. They are also installed in pre-tanked subterranean rooms, as a remedial solution if a high water-table has damaged and broken through the tanked walls and floor.

Superficial measurement: The surface measurement.

Surface-dusting: This refers to concrete floors, such as the concrete oversite-slab to a garage, that – when in use – is continuously producing a cementitious dust. This is believed to be caused by an excessive amount of water in the laid-mix, causing cementitious liquid to rise to the surface.

Survey-appraisal list (externally):   There is no set procedure, but to avoid omissions when surveying a building, the following list is recommended: **1)** Record compass reading of the building's position; **2)** Describe the property briefly (via notes or a Dictaphone); **3)** Check the roof (via binoculars) and include the eaves, gutters and downpipes; **4)** Check the dwelling's walls and carefully note any defects; **5)** Check the upper-floor windows via binoculars, but the ground-floor windows (and doors) close-up; **6)** Check the garage (externally and internally, if possible); **7)** Check the underground drainage, via any accessible inspection covers; **8)** Check the driveway, paving, paths and steps, etc.; **9)** Check the boundary walls and/or fences; **10)** Check the meter cupboards (if fitted externally) for damage; **11)** Check for any trees that might be too close to the building; **12)** Check for alien plant species, such as Japanese knotweed or Giant hogweed.

Survey-appraisal list (internally):   If the vendor of the property is present (and seems to be approachable), ask the following questions: **1)** Do you know what year the property was built? **2)** Have there been any extensions or conversions carried out? **3)** If so, was the work done under Local Building Control Approval? **4)** To your knowledge, has there ever been any blockage-problems with the underground drains? **5)** Has the property ever been rewired? **6)** How old is the boiler? And when was it last serviced? **7)** Where is the rising-main stopcock located? Note that this last question is essential knowledge to pass on to the purchaser, via the Report.

The actual interior survey can either be done by examining and recording the different elements of each room, or by examining and recording each element separately. Either way, the appraisal could be as follows: **1)** Check loft (roof void) or eaves' cupboards; **2a)** Check ceiling(s): their condition and whether they are of plasterboard, lath-and-plaster, Celotex or ACM (asbestos-containing material), etc. (All of which can usually be detected by knuckle-tapping or closed-fist, soft palm-edge light-thumping); Check the condition of cornices or coving; **2b)** Check certain ceiling-areas with a moisture-meter, if deemed necessary via visual discolouration; **3a)** Check the walls; their condition and whether of solid plaster, lath-and-plaster or plasterboard; **3b)** Check all walls with a moisture meter, either randomly or extensively, according to initial findings; **4a)** Check all floors in all rooms, to determine their construction-type and general condition visually – and via heel-thumping on suspended timber-joisted floors; **4b)** Apply a moisture-meter test to any areas of floor suspected of dampness; **5)** Check all radiators for their visual condition and sufficiency of thermostat control valves; **6)** In certain-aged properties, where you would expect to find fireplaces and chimney breasts, keep a lookout for their omission, in the likelihood of discovering a potentially dangerous flying chimney-breast; **7a)** Check room-details generally, i.e., doors, windows, fire-surrounds; **7b)** Check details of boiler type and its location; **7c)** Check the electric fuse box/consumer unit to confirm that MCB (miniature circuit-breaker) switched fuses are fitted; **7d)** Confirm the location of the cold-water mains' stopcock; **7e)** Check for illegal (potentially dangerous) lighting and switches in bath- and/or shower-rooms.

Suspended ceilings and floors: As per British Standards Codes of Practice (BSCP), all ceilings and floors are suspended, but this term generally refers to *false ceilings*, which are either partly- or wholly-separated from the suspended floor above, i.e., if partly separated, they may have their own wall-attachment, but be *hung* in mid-area from the joisted ceiling above. Whereas *suspended floors* are defined as being supported at their ends, without any mid-area support.

Swd: A woodworking trade-abbreviation for **softwood**, the generic term for wood from a coniferous tree.

Swept valley: This traditional valley can be either formed with **plain-tiles** or **slates**, *after* the valley- and cripple-rafters (attached thereto) have been *furred-up* with timber-bearers and a valley board has been laid and fixed to them. This is done to create a transitional bearing to the sharpness of this dihedral-angled turning-point.

With slates, the slate-and-a-half slates are cut to a V-shape (with a flat-portion left at the base of the V) and (to create staggered joints above-and-below the V-shaped slates) other slate-and-a-halves are cut with only one side angled, so as to form a double-width V-shape with a flat portion at their base *when* the two right-angled edges are placed together. So, when these single-width V-shapes and double-width V-shapes are laid and fixed (alternately) one above the other up the valley (with the common oblong slates abutting them on each side), a continuously-slated roof cladding has been achieved around the valley – uninterrupted by the usual indentation of an open-valley *trough* gutter.

And a similar technique is used to form a swept valley with the tile-and-a-half tiles and the common oblong, plain tiles abutting them on each side. Note that if these traditional valleys were used nowadays, a modern-day breathable, impervious membrane (such as Tyvec Supro or Marley Supro) would need to be draped over the rafters, beneath the slating- or tiling-battens and the valley board – as well as over the whole skeletal roof structure.

Swing doors: See **Floor-spring hinge** or **Helical hinges**

System building: This term is a reference to prefabricated, or partly-prefabricated, buildings.

T

T&G: Tongued and grooved.

Tamp/tamped/tamping: In Building, these terms usually refer to the practice of consolidating and levelling newly-laid concrete with the edge of a timber straightedge (of scaffold-board dimensions), which is alternately zig-zagged across the uneven surface, then bumped (tamped) up-and-down, until an even, corrugated surface is achieved.

Tanalized timber: See **Pressure-treated timber**

Tanking: See **Basement rooms and/or cellars**

Tapered gutter: See **Vee roof**

Tapered treads (steps): *Figure 69*:   As per the Building Regulations' Approved Document (AD) K1, the design of tapered treads (historically referred to as *winding steps*) in a staircase must initially be governed by the *theoretical* width of the stairway – not the *actual* width, which would normally be measured from the inner-face of the newel post to the face of the opposite, finished wall-surface. However, when tapered treads are part of a staircase, the theoretical width (unlike the actual width) is related to theoretical pitch lines. To get one's head around this theory, the pitch line could be initially thought of as a taut string line resting on the stair's nosings. So, where it touches the floor at the bottom, it forms a certain pitch-angle, which (for dwelling-house stairs) must not be greater than 42^0. And where it touches the nosing-edges of any two of the parallel steps, it forms theoretical right-angled triangles below the pitch line. So now we have a definable pitch angle on the hypotenuse of the triangle, a measurable base

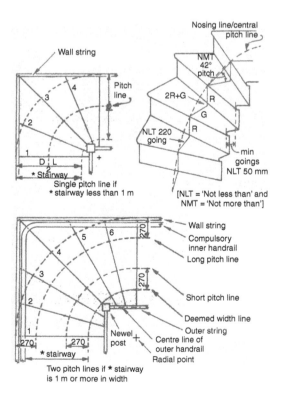

Figure 69 Tapered treads (steps) to the Building Regulations

line (the so-called **going** of the step) and a measurable vertical line (the **rise** of the step). With these last two measurements, a stair designer is then able to test the feasibility of the design by applying them to the Stair Regulations' 2R+G design formula. As quoted elsewhere in this book, this states that twice the *rise* + the *going* must be not less than 550mm/21⅝ in., nor more than 700mm/27½ in.

However, the complex stair-regulations complicate tapered steps further, by stating that when a staircase with tapered steps is *less than 1m (39⅜ in.) wide*, references to 2R+G should be made to one *central* pitch line – but when a stair with tapered steps is *1m (39⅜ in.) or wider*, references to 2R+G should be made to *two* mid-area pitch lines. But the sense of this is clear enough: it creates two different length pitch lines around the quadrant-shaped path of the tapered treads, by which the 2R+G formula can be applied. Hence not less than 550mm of 2R+G for the shorter pitch line and not more than 700mm of 2R+G for the longer pitch line. So, with two pitch lines, one is measured at 270mm (10⅝ in.) from the outer-string's inner newel-post arris to the deemed-width/ *short* pitch line, and the other is also measured at 270mm (10⅝ in.) from the inner wall-string's outer edge to the other deemed-width/*long* pitch line.

Note that the two pitch lines are referred to as being long or short, because they radiate around the tapered treads from a fixed radial point, giving one a shorter radius than the other.

Tap-positioning: Although it may be obvious to most people, it is not always realized that hot and cold-water taps are positioned to suit right-handed people – such being the majority. Hence, the cold-water tap should always be positioned on the right-hand side, thereby giving less risk of accidental scalding to the majority of users. This rule also applies to modern bi-flow taps, via the single lever being turned to the right for cold water and to the left for hot.

Tee hinges: See **Cross-garnet hinges**

Tension and compression areas: *Figure 70*:   The fibres of a loaded beam are subject to a crushing effect referred to as *compressive stress* (compression) across its horizontal surface, below the *topside* – and to being torn apart (stretched) by *tensile stress* (tension) across its horizontal surface, above the *underside*. Being at maximum stress on the top and bottom outer-surfaces, where these two theoretical forces meet in the horizontal *middle* area of a beam, they neutralize and have no stress. This area, thereby, is referred to as the *neutral layer,* or the *neutral axis.*

Terraced houses: A row of houses conjoined to each other by their shared party-walls – half of which, via an imaginary centre-line, is theoretically the legal responsibility of each owner.

Terracotta, hollow building blocks: Made from kiln-fired clay, they have an unglazed or semi-glazed finish, which is of a light-brown orange colour. They are/were used

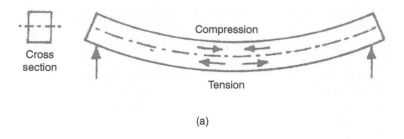

(a)

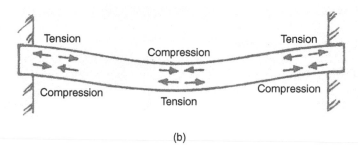

(b)

Figure 70 Tension and compression areas in (a) an unsupported beam (with its ends not built-in) and (b) the same beam with built-in end bearings

for face-work, giving an appearance of a smooth-stonework façade. And in a lesser, standard wall-block size, they were used for partition walls and – with the advantage of their dovetail-indented side-edges – as rows of insulation to the underside of cast-in-situ reinforced-concrete suspended floors. This was achieved by laying them open-end to open-end in beam-spaced rows upon the propped-up floor-shuttering, with specified beam-width spacings between them. When the concrete floor was cast – in between the block-rows and above them by specified dimensions – reinforcing rods were also placed in the tensile area of each beam.

Terrazzo: This is a form of cast-in-situ mosaic flooring, formed by embedding small, random pieces of marble-chippings, coloured glass, etc. (of about 9mm/⅜ in. mesh-size) into a sharp-sand-and-cement floor screed of about 12mm/½ in. thickness. When completely set, the terrazzo is polished with industrial sanders. Note that to avoid thermal cracking, the screed is controlled by being laid within a grid arrangement comprising a honeycomb or square framework of thin, soft metal (such as brass or aluminium). Such floors offer longevity and can be observed in supermarkets.

Thermal bridging: This term is used to describe weak areas in the insulation of a building's envelope, where the thermal values are likely to be lost or weakened.

Thermal expansion and/or contraction: This ubiquitous phenomenon is caused by temperature change, as opposed to movement caused by moisture. It takes place with all materials and forms of construction in building and inevitably damage can be (and often is) caused. For example, it might be thought that a concrete oversite-slab at ground-level formed an inert mass that would not be affected by differences of temperature, but such a thought would be wrong. In fact, it is scientifically known that concrete has a *coefficient of expansion and contraction* equal to steel. So, although heat is generated by the liquified cement upon crystallization, the concrete actually contracts upon setting. Thereby, if the surface upon which the concrete is laid is very rough and uneven – and is not reasonably *smooth* – the newly-laid concrete can be obstructed from *creeping* (contracting) when *setting* takes place. And this can cause multiple cracking to the concrete oversite-slab throughout its thickness, which, upon setting, might be seen on the surface as hairline cracks, looking like a giant-sized jigsaw puzzle.

To avoid such damage, if the substrate is initially comprised of a hardcore base of broken bricks and rubble, it should be overlaid and levelled off with *hogging* (a mixture of sifted gravel and sharp sand. Alternatively, a material used widely nowadays as hogging can be used. It is known as *Mot type 1*, which is comprised of crushed stone aggregates that range from 38mm/1½ in. down to 6mm/¼ in. particles and dust. Once laid and compacted with vibrating machinery or a compacting-plate, the finished substrate meets the required smoothness and density.

Other measures used in industry to combat thermal movement are explained here under the heading of **Expansion joints.** See **Thermal movement (expansion and/or contraction) in buildings**

Thermal/moisture cracks in walls: This refers to cracks initially caused by thermal expansion and contraction of the structure, but which are now aggravated and made worse by rainwater penetration over a period of time.

Thermal movement (expansion and/or contraction) in buildings: Many of the cracks found in either new or old buildings are due to *thermal* movement, caused by changes in temperature, which affects *all* materials. With regard to diagonal cracks seen on the surfaces of walls, either externally or internally, if they are caused by thermal movement (which they invariably are), instead of structural movement, it is usually found that they emanate from the corners of windows or the top corners of doorway openings. And this is usually confirmation of having been caused by thermal movement. Because, when walls expand or contract – as they certainly will – these openings present themselves as nearby areas for the expansion or contraction to manifest itself. Present-day wall-design – where the design permits – often incorporates vertical expansion-joints filled with masonry-mastic to minimise the damage caused by thermal movement. See **Thermal expansion and/or contraction:**

Thermoplastic tiles: Such floor tiles, of 3mm/⅛ in. thickness and in 225mm/9 in. squares, were widely used by contractors for many years on concrete-and-screeded ground floors. They were mainly laid to provide a protective barrier to the bitumen DPM (damp-proof membrane) laid over the sand-and-cement screed – and providing a basic finished-floor was an incidental bonus. It must be mentioned that they are made from thermoplastic resins, etc., and are likely to contain hazardous asbestos fibres if they were laid prior to about 1990. So, taking them up should be avoided for two reasons: **1)** They cannot be removed without breakages, which may release the asbestos fibres, and **2)** Removing them will invariably damage the DPM, which is a barrier against rising damp.

Thermostat: An electrical device for maintaining controlled temperature via a central-heating system within an enclosed structure, such as a dwelling. They can easily be set to a desired temperature between chosen periods of time, and the thermostat will automatically maintain the temperature and switch the boiler on and off at the pro-grammed times.

Three-coat work: Traditionally, this term referred to **1)** So-called *solid plaster*, of about 18mm/¾ in. thickness, on internal wall-surfaces, comprised of **render, float and set** (two coats of sand/cement/lime *rendering*) and a *setting* coat of finishing plaster; **2)** Mastic asphalt (on flat roofs, etc.), applied by a *wooden-float and built-up in three layers;* **3)** Paint applied in three coats, after the priming coat and after all the knots have been coated with knotting (a seemingly much neglected operation nowadays, as seen by the brown, circular stains grinning through).

Threshold/sill: See *Figure 35* on page 84: A hardwood (*floor-plate*) sill at the base of an exterior door, usually attached/jointed to the door jambs.

Throating: See **Anti-capillary grooves**

Thumb latch: A traditional door latch with a D-shaped handle to accommodate one's cupped fingers, and a spoon-handle shaped, pivoted lever above, that when pressed with the thumb, lifts the latch out of the slotted staple fixed to the doorframe's edge. These were – and still are – used on ledge-and-braced, tongue-and-groove boarded doors.

Tie beam: Historically, this was the ceiling-level timber beam of a *roof truss*. It tied together the splay-housed-jointed feet of the opposing *principal rafters*, by holding/tying them into the notched-and-splayed joints with steel straps.

Tie rods: See **Wall-tie bars**

Tight measurements: See **Bare-, full- or tight measurements**

Tile-and-a-half tiles: This description refers to a single, *plain tile* which is 1½ times the width of the common plain tiles being used on a roof. Tile-and-a-halves are used alternately at the gable-end verges to create a staggered-joint bond. As the tiles are 165mm/6½ in. wide x 267mm/10½ in. long – the tile-and-a-halves would be 248mm/9¾ in. wide – and, of course, their lengths would be the same as the ordinary tiles.

Tile and **slate battens:** Pressure-treated softwood battens, graded to BS 5534, are either 50mm/2 in. x 25mm/1 in., or 38mm/1½ in. x 25mm/1 in. Which size to use depends on the spacing of the rafters or roof-trusses and whether relatively light-weight slates are being used or heavier concrete tiles.

Tile-creasing: *Figure 71*: **1)** Sometimes used on low-rise brick boundary-walls, under a brick-on-edge coping, either in one layer of tiles or two layers with centrally-staggered header-joints. The tiles would project on each side of the bricks-on-edge by about 32mm/1¼ in. and – after the bricks had been laid on them – each side-edge would receive a bevelled sand-and-cement fillet; **2)** Also used as a projecting window sill, bedded on the apron-wall brickwork with two staggered layers of tiles laid to an exaggerated weathered-slope away from the window. The tiles should, ideally, be partly under the outer edge of the window and have a damp-proof membrane bedded beneath them.

Tiled-fillet: A right-angled, triangular-shaped sand-and-cement fillet, which is usually regarded as an inferior alternative to stepped-lead flashing against roof-abutments to walls and/or chimney stacks – but the addition of purpose-cut portions of clay- or concrete roof- tiles bedded onto the pre-formed 45° splayed surface of the fillet, is a partly-redeeming feature, as it helps to consolidate the mortar-mass and inhibit surface-erosion. Of course, the tiles do nothing to address the obvious omission of stepped cavity-trays, if the tiled-fillet is against a cavity wall.

Tile-hung/hanging: See **Hung tiles**

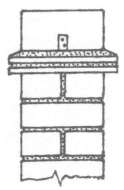

Figure 71 Tile creasing below a brick-on-edge coping to a low-rise wall

Tile/slate hanging: See **Hung tiles**

Tilting fillets: See *Figure 67*: This refers to 75mm x 25mm/3 in. x 1 in. sawn softwood-timber battens with wedge-shaped end-grain sections (having been rip-sawn diagonally along the grain), that are used in a variety of roofing operations, such as being fixed behind the inner, top edge of fascia boards, at the sides of lead-lined valley-boards, and at the top of lead-lined back-gutter boards, etc.

Timber connectors: See **Modern framing anchors**

Timber-grounds: *See Figure 16* on page 34*:* Traditionally, these were used as a base – *ground* – material upon which to nail skirting boards (especially deep, face-stepped, built-up skirtings, requiring a stepped fixing surface) and wall-panelling, etc. For skirtings, the softwood grounds, with a bevelled top-edge to support the eventual rendered-and-set plastering operation, consisted of a sectional size equal to the plaster thickness (say, 16mm/⅝ in.) and a depth of at least 50mm/2 in. Any extra grounds required for deep skirting boards were always fixed vertically (to inhibit *cupping* of the skirting boards), below the horizontal top-ground. And such *soldier-pieces* (uprights), were usually fixed at a maximum of 900mm/3 ft. centres.

Timber saturation point: This refers to seasoned timber used in construction, that has been subjected to too much atmospheric moisture and has reached *saturation point.* This point is not precise, but is reckoned to be between 20% and 30% mc (moisture content). A likely place where this might occur, for example, is in damp, subterranean cellars – especially if the timber joists and their load-bearing wall plates are exposed. Such saturation can promote an attack of wet rot, etc.

Timber-stud partitions: These separating walls are usually non-loadbearing and only of one-storey height. They consist nowadays of *regularized* timber (which is less erratic in its sectional size and shape), and is either of 75mm/3 in. x 50mm/2 in. or 100mm/4 in. x 50mm/2 in. The parts of a stud partition consist of **1)** *two floor plates, one on each side of a door opening*, **2)** a *head plate*, **3)** *two wall studs*, **4)** *two door studs* and a *door head*, **5)** a number of *intermediate studs*, and **6)** a number of *noggings* (which are, in effect, short struts).

Tingles: See **Slate-tingles**

Title deed: A legally written deed/document issued by Land Registry as proof of rightful ownership of a dwelling house or other property in the UK. The original copies of such deeds are held by HM Land Registry Offices – and subsequent changes of ownership are recorded by them, via conveyancing solicitors.

Toggle bolts: This terminology refers to plastic-fittings which are used to create reliable fixings into hollow walls and ceilings. Therefore, the substrate is usually of plasterboard, of which there are two thicknesses: 9.5mm/⅜ in. and 12.5mm/½ in. However, they can be used for fixings into any other relatively thin lining material. With the introduction of dry-lined walls more than four decades ago – which have now mostly superseded newly-formed solid-plastered walls – there has been more demand for reliable fixings. And manufacturers have risen to this with a wide variety of devices now on the market. Only five of them are named here: **1)** Plasplugs Super Toggle Cavity Anchors; **2)** Plasplugs Plasterboard Heavy-Duty Fixings; **3)** Fischer Plasterboard Plugs; **4)** Nylon or Metal (zinc-plated) Easi-Drivers; and **5)** Rawlplug Uno Plugs.

Tongue-and-grooved (T&G) flooring: This type of boarding has always been popular as a good quality subfloor on suspended timber joists. However, in recent decades – for reasons of cost – it has been largely superseded by T&G panels of flooring-quality chipboard, plywood, or Sterling OSB (oriented strand board).

Tonk strips: These are used for simply-supported shelves that are to be contained within a bookcase with – at least – side cheeks. They consist of vertical metal strips with pierced slots to support the small, metal lugs that act as interchangeable bracket-bearers, giving infinitely variable shelf-spacing options. Professionally, Tonk strips are housed into vertical, shallow grooves in the side cheeks, but simpler, face-fixing types are available.

Toothed timber-connectors: See *Figure 74* on page 254: These 50mm/63mm or 75mm diameter 'Bulldog' connectors are used between structural timbers (such as building up the width of a timber-beam, etc.) via 12mm diameter bolts and surface-washers.

Toothing and block-bonding: These are traditional bricklaying techniques that can be used in joining new brickwork (or blockwork) onto (and into) existing walls. Nowadays, this seems to be mostly done by coach-screwing *stainless-steel wall-starter kits* to the wall, as described here under **Wall-starter ties**.

Toothing relates to carefully creating or cutting brick-height indents in a wall, rising up vertically in a hit-and-miss pattern, to enable a wall to be tooth-bonded to it – thereby continuing the bond-pattern and avoiding a continuous, straight abutment joint. Block-bonding is a similar technique used for the same reasons, except that – whether it be built with brickwork or blockwork – the toothed hit-and-miss pattern is formed to suit either three-courses of brickwork or the height of a building block (which equals three-courses of brickwork) being cut into the recipient wall in the hit-and-miss pattern.

These traditional techniques, although more time-consuming, are arguably more attractive than a completely vertical abutment-joint created with wall-starter kits. However, vertical abutment-joints are also now becoming common as *vertical expansion-joints*.

Top-hung casement windows: Fanlights and other top-hung windows that must open outwards for weathering reasons.

Torched-on roofing felt: See *Figure 80* on page 269: This alternative to traditional flat-roof coverings still uses three layers of (*modified*) bituminized felt, but the main differences are: **1)** the technique of *laying and bonding* has changed from brushed-on hot-bonding, to torched-on self-bonding and, **2)** the modified top layer of felt has a close-knit mineralized finish (in various optional colours) and is approximately 4mm thick. The first layer is laid/bonded onto a primed deck, then the second (sanded) underlay is heated on its underside (torched-on) whilst being rolled out over the first layer. The third (top layer) – which is a fire-rated polyester-based, mineral-coated felt, available in various colours – is also torched on. All layers must be overlapped at their edges, as well as (importantly) being laid at right-angles to each other. Finally, fore-thought must be used to end up with final overlaps pointing towards (and being parallel to) the pre-formed mineral-faced drip-edge into the gutter. But this is easy enough, because (with three layers), if the first is laid parallel to the drip-edge into the gutter, the third layer will end up in the same direction.

Torching: *Figure 72:*   This term refers to a 19th-century technique that was used to stop wind-driven rain from being driven *uphill,* under the interconnecting edges of plain-tile-clad roof-surfaces. After the roof-slopes had been battened and tiled, the underside top-edges of all the tiles (which invariably included the top edges of the tile-battens), were tediously and roughly pointed (torched) with a mortar mixture of sand, cement and lime. This technique was used because, prior to the early 1930s, slated or tiled pitched-roofs did not have the present-day substrate protection of an impervious roofing-felt underlay draped over their rafters, below the slate- or tile-battens. Up until this date, it did not yet exist.

   To conclude, torching helped to hold back the driven rain, but it deteriorated and broke down over time; ending up, after being blown, as a heavy film of mortar-dust covering the surfaces of a loft's ceiling joists and lath-and-plaster ceilings – and leaving

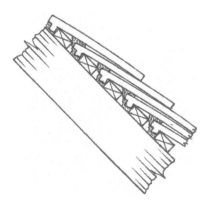

Figure 72 Mortar-torching between plain tiles without any roofing underfelt below tile battens

the untreated tile-battens in a wet-rotted condition. There is evidence of a number of these anciently-clad roofs still being in existence. Hence the essential need of a building surveyor, acting for a purchaser (or a Building Society), to always examine any accessible roof, if possible.

Total going (TG) and total rise (TR): *Figure 73:*   In the basic theory of stair-design, one movement of a person's foot *going* forward is referred to as 'the *going*' of the step and the other foot *rising* up (to another level) is referred to as 'the *rise*' of the step. Therefore, from a design point-of-view (at the drawing-board stage), all of the step-movements going forward are referred to as 'the *total going*' *(TG)* and all of the step-movements rising up are referred to as 'the *total rise*' *(TR)*. These two terms (TG and TR) are important references to the simple maths and geometrical division required to design flights of stairs that meet the sometimes-complex requirements of *AD K1* of the Building Regulations.

Toughened glass: See **Safety glass**

Tower bolts: Obtainable in various lengths, the sliding bolt is encased by a strong, open- metalwork casing. The longer tower bolts are sometimes used for heavy, industrial doors and may have a long monkey-tail, cranked-end projection, to assist in the use of the bolt on tall doors. The *holding* part of the bolt is called a staple.

TRADA: These initials are an acronym for Timber Research and Development Association, which is an autonomous body set up by the government many years ago.

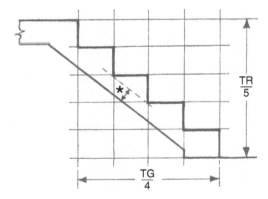

Figure 73 Stair design is based on the division of the 'total going' (TG) of a stair flight, in relation
to the 'total rise' (TR). In the above example, 5 risers automatically produce 4 goings
(because the top landing-step is always excluded). *This measurement on concrete-
stairs is referred to as the 'waist thickness'

TRADA-recommended joist-sizes: These details are available from this association, under the title of *Span tables for solid timber members in floors, ceilings and roofs for dwellings*. Their contact details are: Bookshop at http://bookshop.trada.co.uk. Note that prior to the early years of the 21st century, detailed span tables were freely printed in copies of The Building Regulations.

TRADA roof trusses: *Figure 74*:   Soon after WWII, in an exercise to economise on timber-sizes in roofing systems, TRADA (often abbreviated to TDA) designed a new system of roofing that used newly-designed roof-trusses, to be spaced at 1.8m/ 5ft.11in. apart to support the lighter-weight purlins and the normal-sized binders at ceiling-joist level. Once the trusses were erected and braced, the roof-areas between the trusses were infilled with lightweight common rafters of a 75mm x 50mm/3in. x 2in. section. The overlapped and sandwiched joints of the trusses were usually bolted together onsite with 12mm/½ in. diameter bolts and 63mm/2½ in. diameter toothed Bulldog connectors. Such trusses as these were the forerunners of modern-day roof trusses, using even lighter-weight, factory-made trusses and underside board-bracing in lieu of heavy purlins and ridge boards.

Trammel: Used for setting out circular-work, or semi-elliptical arches, etc. They are usually in the form of *trammel-heads* or *beam-compass heads* that can be fitted to a length of batten.

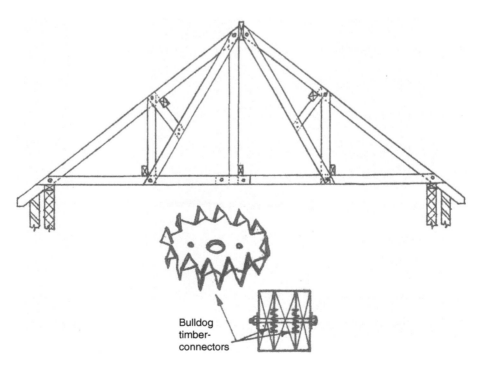

Figure 74 TRADA (Timber Research and Development Association) roof truss, bolted together on site with 63mmϕ, toothed 'Bulldog' timber-connectors

Transom: A horizontal window-frame component, usually of wood, stone or metal, that separates the lower casements from the upper (usually smaller) windows.

Traps: See **S-traps** and **P-traps**

Treads: The horizontal surfaces of stairs/steps.

Treated timber: This term usually refers to structural building-timbers that have been treated with chemical preservatives in pressurized vats. Two of these well-known procedures are tanalizing and protimizing.

Tree Preservation Order (TPO):   This is an Order made by a local planning authority in England to protect specific trees, groups of trees or woodlands in the interests of amenity. An Order prohibits the cutting down, topping, lopping, uprooting, wilful damage and wilful destruction of trees without a local planning authority's written consent. If consent is given, it can be subject to conditions which have to be followed. Cutting roots is also a prohibited activity and requires the authority's consent. Protected trees can be of any size or species. This information was gleaned from a gov.uk portal on TPOs.

Tree-root damage to dwellings: Trees found growing too near to a building, in areas known to be of shrinkable-clay, could cause settlement or subsidence damage to the foundations by their roots removing too much water from the clay. In this respect, the species of the tree(s) should be identified, if possible, as some consume more water from the ground than others. If tree-threats are discovered during a building survey, a recommendation should be made for either a periodic appraisal of the nearby wall's condition, or an imminent subsoil test and appraisal by an arboriculturist. Note that cutting down and removing trees in areas of shrinkable clay can also cause foundation-damage known as **heave**, whereby the subsoil swells and lifts the foundations at less-loaded points of the superstructure, such as under low-rise apron walls below window-openings. British Standard BS 5837:1980: Code of Practice for Trees in Relation to Construction, states that: "Where the presence of shrinkable clay has been established and where roots have been found or are anticipated in the future, it is advisable to take precautionary measures in the design of the building's foundations (in accordance with CP 2004:1972 and CP 101:1972). Alternatively, consideration should be given to designing the superstructure to accommodate any foundation movement that might be induced by clay shrinkage. Where a subsoil investigation has not been carried out, an approximate rule-of-thumb guide is that on shrinkable clay, if risk of damage is to be minimized, special precautionary measures with foundations or superstructure should be considered where the height or anticipated height of the tree exceeds its distance from the building."

With all due respect to BS 5837:1980, it does not address tree-root damage to existing buildings and their foundations. It unambiguously covers and recommends taking necessary precautionary measures in the *design of a new building's foundations* in relation to trees.

However, the suggestion of an approximate rule-of-thumb guide regarding the height of trees in relation to the distance from *newly-designed* foundations is very useful – and can also be used when trees are found to violate this rule in relation to mature properties built on *undesigned* strip foundations, etc.

Trench-fill foundations: See **Strip foundations** and **Trench-fill foundations**

Trench protection systems: Traditionally, this procedure was referred to as *timbering to trenches*, which referred to a variety of lateral-propping arrangements to support the sides of freshly-dug trenches during work – such as drain-laying – being carried out. In its simplest form, this involved closely-spaced, vertical *poling boards* (with a 40 x 225mm/1½ x 9in sectional size), horizontal *waling* timbers (with a 100 x 150mm/4in x 6in sectional size) that supported the poling boards – and 100 x 100mm/4in x 4in timber *struts* that supported the waling timbers placed on each side of the trench. Note that timbering to trenches – in more complex, illustrated arrangements – can be found in such books as *Book 3, Carpentry & Joinery* by R. Bayliss, Hutchinson Educational Ltd., or other yesteryear technical books.

Note that injuries and death by unsupported trench-sides collapsing whilst workers are in the trenches was quite common historically – and isolated incidences of death via unsupported trenches are still reported upon in some countries.

Trickle vents: These are slim, longitudinal plastic-fittings, usually fitted to the top rails of uPVC or wooden casement window-frame heads, to allow optional (yet essential) room-ventilation when the window is closed, via a slide-button or lever.

Trim: This is an American/Canadian building-term for architrave and/or skirting-boards.

Trimmed joists: See **Floor joists**

Trimmed rafters: The technical nomenclature for roofing rafters is very similar to those used for floor joists. Therefore, *trimmed* joists (which are joined at right-angles to *trimming* joists) act like *trimmed rafters* joined to *trimming rafters*. As an example, the pitched rafters on each side of a dormer window are trimming rafters (so, should be thicker, or doubled-up, side-by-side); and where the dormer's ceilinged-area (and its windowed façade) meets the pitched roof, these will be supported by trimmer rafters. And these open areas will be infilled with trimmed rafters and studs.

Trimmer arch: This refers to a part-vaulted brick-arch that was built traditionally within the 225mm/9 in. depth of a suspended timber-joisted floor. Its purpose was to support a sand-and-cement screeded fireplace hearth – so the partly-segmental brick-on-edge arch was built within the confined depth of the trimming- and trimmer-joists of the fireplace! It sprang from the front-side of the chimney-breast wall to within 25mm/ 1 in. of the top of the trimmer joist (to allow for the sand-and-cement screed); on the underside of the arch, a few shaped, timber firring pieces were fixed at right-angles to the trimmer, for the fixing of the ceiling-laths.

Trimmer joists: See **Floor joists**

Trimmer rafters: See **Trimmed rafters**

Trimming joists: See **Floor joists**

Trimming rafters: See **Trimmed rafters**

Trough gutter: See **Box gutters**

Trussed and/or diagonally braced partitions: Such timber-framed partitions have not been built for many decades now, but builders or DIYers doing conversion work on old-build properties need to be aware of altering or removing a trussed partition. It was built to carry its own weight *and* the weight of the floor above. This was achieved by having a timber (beam-like) head-plate of 200mm/8 in. depth, an *intertie* beam above the door-head plate of a similar depth and two pairs of diagonal bracing studs on each side of the mid-area doorway (one above the other), each in an arrowhead formation towards the top-beam's centre. Finally, there were two very long, 25mm/1 in. diameter tie bolts fixed vertically (through drilled holes) within the partition, on each side of the door posts/studs.

Truss-rafter advantages: One of the advantages of truss-rafter roofs is the clear span that they achieve, without the traditional-need for load-bearing partitions or walls in the mid-span area. Another advantage to the building designer/architect is that the truss-fabricator will only need basic architectural-design information to plan the truss layout in detail for the building-designer's approval.

Truss-rafter bracing: See *Figure 39* on page 101: Modern truss-rafter assemblies – as opposed to traditional cut-and-pitched roofs – must be permanently braced (on their undersides) strictly to the designer's specification. And the bracing-arrangement details given below are based on information in the technical manuals obtained from Gang-Nail Systems Ltd, a member company of the International Truss Plate Association:

1)   Temporary longitudinal bracing, used to stabilize the position of the trusses during erection, should be of a 75mm x 25mm, or a 100mm x 25mm section.

2) Permanent diagonal bracing, forming 45^0 angles to the underside of the rafters, should be of a 97mm x 22mm minimum section and should run from the highest point on the underside of a truss to overlap and fix to the wall plate, starting at a gable end and running zigzag throughout the length of the roof, fixed to every truss with two 3.35mm x 65mm-long galvanised wire nails. There should be not less than four braces (two on each slope) of any short-length duo-pitched roof. All joins of incomplete bracing-lengths should be overlapped (side-by-side) by at least two trussed rafters. Note that the angle-of-bracing given above as ideally 45^0, should not be less than 35^0 or more than 50^0.

3) Permanent, longitudinal bracing should be of 97mm x 22mm minimum section, with fixings and overlap allowances as for diagonal bracing, positioned at all *node points* (points on a truss where members intersect), with a 25mm offset from the underside of the rafters (*top chords*) to clear the diagonal bracing, extended through the whole length of the roof – and butted tight up against party- or gable-walls

4) Permanent, diagonal-web chevron-bracing, should also be of 97mm x 22mm minimum section – and each diagonal should be extended over at least three trusses if required for duo-pitched spans over 8m., or for mono-pitched spans over 5m.

On pre-sale building inspections of a dwelling and its roof-void (via the trap-hatch), this essential bracing, with regard to its layout, is occasionally found to have parts missing – or the overlapping of the bracing members may have been skimped. Such a discovery requires to be reported upon for remedial work to be carried out.

Truss-rafter roofs: This term usually applies to modern-day truss-rafter assemblies that are factory-designed, factory-made and that are delivered to site for site-assembly by carpenters.

Tuck pointing: See *Figure 59* on page 191: A traditional method of smartening up old face-brickwork. The mortar joints were raked out and refilled *flush* with a coloured mix similar to the brickwork-colour, by using red sand or brick-making dust. Whilst unset, the joints were indented with a *jointing tool* and filled with a white mortar-mix containing lime-putty and silver sand. The protruding white pointing was applied by using a mortar-loaded jointing tool, guided by a bricklayer's *pointing rule* (a short straightedge). Finally, the ragged edges of the protruding tuck pointing were cut off at the top and bottom by running a bent-tipped knife – a so-called *Frenchman* – along the edges of the bricklayer's *pointing rule* (a miniature straightedge), with a rubber pad at each end of the inner-face, to allow the mortar droppings to fall clear.

Tudor and **mock-Tudor façades:**   There are still a certain number of historical houses around that date back to the Tudor period of the 15th to the early 17th century, that were built with English oak-framed timber beams, posts, diagonal braces and studs, etc., wattle-and-daub panels – or (in the latter part of the period) – half-brick thick (4½ in./115mm) infilled panels, laid to stretcher bond or herringbone-pattern bond. Note that this thickness of single-*leaf* outer-wall would contravene the Building Regulations nowadays – but changes in these regulations cannot be applied/enforced retrospectively.

Mock-Tudor façades nowadays emulate the original timber-framework and offer the advantages of being superimposed on modern cavity-walls. Various thicknesses and widths of boards – imitating the beams, posts and studs (not usually the diagonal braces) – can be seen fixed to the brick- or rendered-façades. The timbering is usually dark-oak stained and of a sawn finish (as opposed to planed). It can be either, but it should be pressure-treated with preservative, otherwise it will likely have a short lifespan. Or, Accoya® wood may have been used, with a 50-year warranted use above ground and without the need for carbon-producing, pressurized preservatives. Note that if one finds these mock-timber-framed façades difficult to distinguish from the historical originals, then maybe the architect and builder have done their jobs well.

Tumbling-in courses: This term refers to tilted courses of internal or external brick-work used to create an inclined, sloping surface – as opposed to flat-topped *corbelled* (stepped) surfaces. As an example, such work is used externally when a chimney breast – containing widely separated flues at its base – is being reduced to a narrower width to create the necessary transition from a chimney breast to a chimney stack. The tumbled-in outer bricks, thereby create a weathered slope – as opposed to a stepped slope.

Turning piece: A basic form of a traditional *arch centre*, made from a solid piece of timber (perhaps with a 100 x 100mm/4 in. x 4 in. sectional size), with a segmental shape on its topside to hold arch-bricks (voussoirs) up to a 75mm/3 in. segmental rise.

Twist: 1) Warping that can occur in timber after seasoning and conversion (to sawn sizes), which produces distortion in length to a spiral-like propeller-shape. 2) Distortion in a framed-up unit, caused by one or more of the components being twisted, or by ill-formed or ill-fitting corner joints.

Two-coat plaster: Nowadays, so-called *solid plaster*, when used on internal solid walls, is comprised of a rendering-coat and a setting-coat, commonly referred to as *render-and-set.*

Two-pipe system: Apart from a heating circuit for central-heating, this term refers to a *soil-and-waste* system comprised of two separate pipes; one being a large-diameter *soil pipe* to take the WC waste to the underground drain, the other being a 40mm/1½ in. diameter *waste pipe* to take the waste-water from wash-hand basins and sink to the drainage system, via an exterior trapped-gulley.

U

Undercarriages of **staircases:** See **Staircase undercarriages** and/or **Brackets**

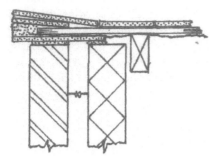

Figure 75 Plain-tiled verge of a pitched roof, showing undercloak tiles below the tile battens and the roofing felt

Undercloak tiles, slates or fibre-cement strips: *Figure 75*:   This is a course (row) of plain tiles, slates, or fibre-cement-board strips, fixed or bedded at the verges of a roof, to give support (and adhesion) to the overhanging verge-tiles or slates. The upper-tiles or slates of the verges are bedded onto the undercloak tiles, slates or fibre strip with sand-and-cement mortar. Nowadays, the fibre strips (which often contained chrysotile asbestos fibres) have been superseded by asbestos-free sheet material such as Cembrit undercloak strips of varying widths; 150mm/6in. widths seemingly being the most commonly used.

Under-eaves tiles/or slates and top-course tiles/or slates: When 'plain' tiles (or slates) are used, it is necessary to start at the eaves with a row of under-eaves tiles or slates – and to finish at the top with a row of *top-course tiles* or *slates* (both of which are shorter in length). This is to eliminate the open joints that would otherwise occur between the abutting plain-tiles and/or slates on the first- and the last-laid rows. The plain 'under-eaves' tiles are obtainable in ready-made shorter lengths (which can also be used for the top-course tiles), but quarried slates and fibre-cement slates have to be reduced/hand-cut on site.

Underfloor (radiant) heating: This operates by infrared energy, i.e., thermal radiation principally via **1)** coiled panels of smallbore, plastic or metal waterpipes laid on rigid insulation and overlaid with a 65mm/2½ in. sand-and-cement floor screed; or **2)** electric underfloor heating mats, laid on rigid insulation and used under a variety of finished, floor materials.

Underground-drainage systems from buildings: Depending on sewage arrangements that may vary in different districts, drainage systems may be either **1)** *One-pipe systems*, that receive so-called *foul water* from WCs, *soiled water* from wash-hand basins and sinks, plus rainwater from roofs and surface-gullies; or **2)** *Two-pipe systems*, that keep the rainwater separate from the foul/soiled water (to relieve the sewage-treatment works). Both systems have at least one inspection chamber (manhole), depending on the length of drainage before it discharges into the public sewer and/or the number of bends involved en route to the sewer. For 100mm/4 in. diameter pipes, used for one or two households, the usual gradient/rate of fall is 1 in 40.

Underpinning: When any building-structure is analysed by a surveyor as being very likely to have suffered subsidence, this is usually based entirely on the evidence of defects seen in the above-ground walls (externally and sometimes internally). And the advice to the property-owner is to inform their insurers, because the latter need independent confirmation from their own surveyor.

If remedial work is sanctioned and underpinning is to take place – there are many new techniques available nowadays, which are less intrusive and less labour-intensive than the old system of forming piecemeal removal-and-reinstatement of hit-and-miss portions of new concrete-foundation and toothed-wall portions every 1.5m/4ft.11in. maximum, below the old foundations – and then waiting for them to set before removing and renewing the remaining old portions of wall onto new, reinforced-linked portions of foundation. Also, with this old method, a designed system of raking shores was usually required, to support the building.

One of the new techniques involves a partial spot-removal of the interior ground-floor structure (near the outer walls), mechanical excavation and insertion of short-length concrete piles. Then, opposite these piles, more are inserted on the exterior of the building (again, near the outer walls) and short-length, precast reinforced-concrete beams (needles) are placed on the tops of the piles to permanently relieve the foundations and support the walls.

Under-ridge tiles: These apply only to so-called *plain tiles*, which require shorter-length tiles to be laid in the top tiling-course of a roof, below the semi-circular or segmental-shaped ridge tiles. Under-ridge tiles are 225mm/9 in. long, instead of the 270mm/10⅝ in. length for standard plain tiles.

Undressed or **unwrought timber:** Un-planed, sawn timber.

Unequal settlement: See **Settlement appraisal (internally and externally)**

Unplasticized polyvinyl chloride (uPVC): In its rigid – and pliable (plasticized) form – this almost non-combustible material has transformed many building materials for more than five decades. UPVC casement windows, doorframes with doors, fascia-board and bargeboard sections (for covering their wooden counterparts, or replacing them) are ubiquitous in the UK nowadays. As are plumbing items such as soil pipes, waste pipes, drain pipes, gutters and downpipes, etc., and electrical items, such as plastic conduit pipes and trunking, socket outlets and light switches, as well as plasticized PVC, in the form of plastic-sheathed cables.

UPVC windows and doors: As mentioned above, these uPVC products are very popular and seem to have almost taken over the market from their wooden counterparts. The latter mainly losing the battle via their need for (and most people's neglect of) regular, four- to five-yearly maintenance/repainting or re-staining, etc. Maybe the fairly recent introduction of **Accoya® wood**, which has a warranted 50-year lifespan in above-ground use (and a 25-year lifespan in below-ground use) will encourage a wooden-window and door revival. However, when appraising uPVC windows and/or doors (especially doors), check whether they open, close and lock properly, take note of how vibratory the frames are, because such are the signs of cheaper units, made without internal metal-framing.

Utilities: This is a common reference to available services such as electricity, gas, sewage disposal and water, etc.

U-values: Stated U-values refer to the thermal transmittance of heat through a material or a construction, such as a wall, floor or roof, etc.

V

Valley: Where two sloping (pitched) surfaces meet and form an inverted dihedral angle.

Valley boards: *Figure 76***:**   If a roof-valley (where two pitched roofs intersect at right angles) is to be traditionally lead-lined, then two valley boards are usually fitted into the recess, laid and fixed side by side. The sawn boards should be 25mm/1 in. thick x 225mm/9 in. wide. And a *tilting fillet* (of ex 75mm/3 in. x 25mm/1 in.) should be fixed at the top edge of each board, ready to be included in the lining of the valley with sheet lead, etc.

Valley gutter: The impervious weathering of a small width of the sloping (pitched) surfaces that are formed by the meeting of two pitched roofs. The impervious material is traditionally sheet lead (of a good weight), but may also be of fibreglass, etc. See **Valley junctions.**

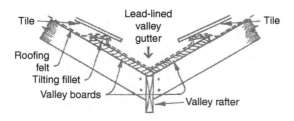

Figure 76 Valley boards and valley rafters

Valley junctions: See herein under title headings: **1)** GRP (glass-reinforced plastic) valley liners; **2)** Hogsback-tiled valley; **3)** Laced valley; **4)** Mitred valley (with cut-tiles and lead-soakers); **5)** Open-valley with lead-lined gutter; **6)** Purpose-made tiles' valley; **7)** Swept valley.

Valley rafters: See *Figure 76* on page 262: These are pairs of rafters, diminishing like hip rafters, spanning from the ridge boards to the inverted valley rafter – historically called *cripple rafters*, via having their usual birdsmouthed-and-plumb-cut *feet* cut off.

Valley tiles: These are specially made to bond with the abutting edges of *plain tiles* (not *interlocking tiles*) and they are used instead of *open trough valleys* (which are shallow, lead-lined, zinc-lined or fibreglass-lined recesses/troughs that are more commonly used for valleys nowadays – especially the fibreglass valley-liners. The valley tiles are odd-looking V-shaped tiles with concaved upper-surfaces and the lower portion of the V missing.

Valuers of property: See **Property valuation**

Valve: A mechanical device to close, open, or regulate a flow of water or gas in a pipeline or system. In its simplest form, it is often referred to as a stopcock. In its more modern form, it would be fitted in a dual push-button, valve-flushing WC cistern, to regulate the flow.

Vanity unit: A cupboard unit fixed to a wall, with a wash-hand basin fitted into its top.

Vapour barrier: This is sometimes referred to as a *moisture barrier*. Its function is to stop the diffusion of moisture-movement through ceilings, floors or walls to the dwelling's surface-material, via *interstitial condensation*. Such materials used as vapour barriers are usually of plastic, silver foil or polyethylene.

Variation Orders: An important written notice – usually written and issued by an architect visiting a building site – authorizing the builder/building contractor to carry out extra work. Such additional work usually emanates from practical, unforeseen issues that the contractor has brought to the architect's notice – or that the architect requests, because he wants to vary from the contract drawings or specification for various (usually practical) reasons. Variation Orders are not issued lightly, because a contractor will be compiling daywork labour-rates, materials and profit against each order and presenting the architect/client with an extra bill.

Vaulted ceiling: A semi-circular arched ceiling, usually of bonded brickwork, that springs from brickwork side-walls. The vaulted structure, whilst providing an

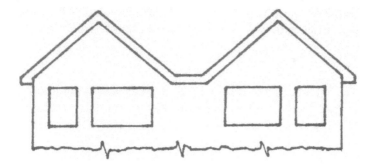

Figure 77 Vee roof with a mid-area tapered gutter

underground room, a tunnel or cellar, etc., provides excellent support for the backfilled ground or floor above.

Vee joint: A 6mm/¼ in. wide 45⁰ chamfer on the two face-side edges of tongue-and-grooved matchboarding, to provide a decorative feature, whilst masking ubiquitous shrinkage.

Vee roof: *Figure 77*:   Two pitched roofs – usually gable-ended – that meet on their side flanks and form a vee-shaped valley between them. The sharpness of the vee formation has to be flattened out at its base to create a *tapered gutter*, which must be wider at one end to create a *fall*. This is done with short *firring* joists and tapered valley-boards. At each side of the boarded gutter-base, more boards with splayed lower-edges are fixed to the eaves' rafters to form the sides of the gutter. The three boarded surfaces now described that make up the flat-based vee gutter are of 225mm/9 in. x 25mm/1 in. sawn timber. And on each of the sloping sideboards' top-surfaces, important *tilting fillets* are fixed. These are *dressed* over with the lead-lining that covers the tapered gutter, ready to carry the roofing fabric and the first tiles or slates that project over them on each side of the tapered gutter.

Verdigris: A green or blue acetate formed as a protective patina on copper exposed to the air.

Verge: See *Figure 9* on page 18: The outer, side-edge of a sloping (pitched) roof which overhangs the gable wall. For weathering purposes, the overhang was established traditionally as 40mm to 50mm/1½ in. to 2 in.

Verge tiles or **slates:**   See **Tile-and-a-half tiles** or **Slate-and-a-half slates**

Vertical tiling: See **Hung tiles**

Vh: An abbreviation for *vertical height*, as sometimes used on a building drawing.

Vinyl floor-tiles: In the mid-1900s, up until the late 2000s, such tiles (containing asbestos fibres) were used widely on newbuild-properties at ground-floor level, as a means of sealing and protecting the sand-and-cement floor screeds. The bitumen compound, upon which many of these tiles were laid, also acted as a DPM (damp-proof membrane).

VOCs (Volatile Organic Compounds):   Such chemicals have a high vapour and a low water-solubility. And they are known to emanate from substances used in the manufacture of certain paint mediums, varnishes, strippers, wood preservatives, etc. Some VOCs are known to be dangerous to health and/or cause harm to the environment. Most VOCs, apparently, are not acutely toxic, but may produce chronic health issues.

Voussoirs: Tapered bricks in a gauged brick arch.

W

Wagtail: A relatively modern term for a slim, narrow section of timber that hangs securely but loosely (like a pendulum) within the side-boxes of a boxframe window, to keep the two sash-weights therein from clashing and jamming, as either the top sash is lowered or the bottom sash is raised.

Wainscot/wainscotting: These archaic terms refer to *skirting boards*; although they also seem to have been used archaically in reference to *wall-panelling*.

Waist thickness: See *Figure 73* on page 253: The lower portion of a concrete stair, acting as an inclined reinforced slab, whereby its structural-thickness is measured at right-angles to the sloping soffit, up to the point where the inverted angles of the treads and risers meet.

Waling(s):   This term refers to longitudinal side-alignment timbers used in temporary, panelled formwork when forming reinforced-concrete walls.

Wall-panelling: See **Panelled walls**

Wall-plastering: See **Plastered (interior) walls**

Wall-plate and gable-end restraint-straps: See *Figure 2* on page 6 and/or *Figure 51* on page 124: These are screw-hole perforated galvanized-steel, L-shaped straps, introduced by Building Regulations after a very destructive storm in late 1987, to restrain a roof's gable-ended walls and roof from severe wind-damage. The hooked straps hold down the roof via their multi-screwed connection to its timber wall-plates and to the inner-surfaces of the walls below. Other restraint straps, in the roof void, hold onto the gable walls by being fixed at right-angles to the undersides of the rafters or roof-trusses.

Wall plates: This term is used for pressure-treated timber bearers, usually with a sectional-size of at least 100mm/4 in. x 50mm/2 in., that are still used as seating/fixing bearers below timber, pitched and flat-roofs, or – perhaps very rarely nowadays – as bearers for suspended timber-joists at ground-floor level. In all cases, they would be mortar-bedded to the substrate – and in the latter case, a damp-proof membrane would need to be sandwiched in the mortar.

Wall plugs: Plastic wall-plugs have been the foremost screw-fixing device for many years now (at least four decades) and are very reliable if used properly when fixing timber or joinery items, etc., to solid walls. They are either made of high-density poly-thene or nylon and are usually colour-coded in relation to their screw-gauge size. Some wall plugs have small, flexible barbs projecting from each side, like the head of an arrow. This helps to create an initial anchorage in the plugged wall, thus allowing the screw to be turned without the negative effect of rotating the plug.

Wall-starter ties: *Figure 78*:   These metal components are a modern alternative to traditional *toothing* and *block-bonding* techniques used for joining newly-built walls onto existing walls. As an example of their use, if a single-storey extension (with double-leaf cavity walls) was to be built up against the wall of an existing dwelling, the first pair (one for each leaf) of stainless-steel wall-starter channels (of 1.2m/4 ft. length) would be vertically coach-screwed to the existing walls and the new brickwork (for the outer leaf) and blockwork (for the inner leaf) would be bedded up against the channels. Then (essentially) special wall-ties (supplied with the kit) are inserted,

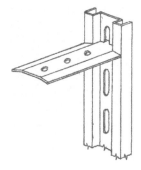

Figure 78 Stainless-steel tie shown in position on a patented Wall Starter Kit fixed to a wall

twisted and locked into the channels and bedded onto the first level of bricks and blocks. This process is repeated at every three- courses of brickwork and every one course of blockwork rise, i.e., every 225mm/9 in. Note that this attachment to the wall will create a narrow, vertical abutment joint (similar to a modern-day expansion joint) which should eventually be gunned with a mastic sealant. To understand the age-old traditional method, mentioned above, see the **Toothing and block-bonding** heading in this book.

Wall string: See **Inner string**

Wall ties: See *Figure 20* on page 48: Lightweight, metal *bridging*-connectors, used to tie the inner- and outer-skins of cavity walls together, as the walls are being built. See detailed procedure above under the **Cavity walls** heading – or below, under the **Wall-tie corrosion and expansion** heading.

Wall-tie bars: *Figure 79*: Buildings aged up to about 1900 can occasionally be found displaying at least two large **S-**, **X-** or **Y-ended**, steel wall-tie shapes positioned on the two opposing sides of the building's elevations. And in the centre of the S-, X- or the Y-shaped restraint plates, the ends of the threaded tie-bars, nuts and washers will also be seen. The steel tie-bars are usually at least 18mm/¾ in. diameter and they were used as remedial treatment for structural defects such as bulging and/or leaning walls, or where unrestrained gable-end walls were showing signs of movement. Inside such properties, efforts were always made to conceal the tie bars. They were either positioned within the confines of a floor-construction, parallel to the joists, or notched across the tops of the joists, below the floor boards. Restraining the opposing gable-ended walls was usually more easily solved, by creating a number of timber collars across the opposing pitched-rafters and laying the wall-tie bar across their mid-area.

Wall-tie corrosion and expansion: See *Figure 20* on page 48: Over many decades now, zinc-plated or galvanized mild-steel wall-ties have been used to hold the two *skins* of

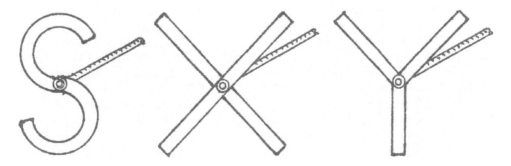

Figure 79 S-, X- and Y-ended wall-tie bars

cavity-walls together. But, because the outer skins of brickwork soak up a certain amount of driven rain, this can affect the wall ties, which may corrode, causing a build-up of iron oxide (also known as rust). Over time, this causes the flat-shaped-type of ties to expand, via layer-upon-layer of rust. Eventually, this pressure within the horizontal mortar joints results in ever-expanding bed-joint cracks. And if this is neglected for too long, the wall can start to bulge and, eventually, areas of outer-wall (large or small) will start to collapse.

The first signs of such an occurrence – usually only detectable in its early stages by a suspicious surveyor's close scrutiny – are seen to be a hit-and-miss irregular pattern of fine cracks in the mortar-bed joints, usually at six brick-courses (450mm/18 in.) apart vertically.

If building surveyors or other trade-knowledgeable people raise awareness of these cracks, it would be wise to pay a small fee for an independent *wall-tie endoscopy* survey to be carried out. This can be done by a specialist or a CSRT surveyor (Certified Surveyor of Remedial Treatment). And if wall-tie corrosion is confirmed, then quotes can be sought for near-future wall-tie replacement treatment. Note, however, that if a surveyor suspects wall-tie corrosion on a property being surveyed for a prospective purchaser, such an invasive technique as wall-tie endoscopy (drilling one or two 6mm/¼ in. diameter holes in the mortar joints, to provide access of the cavity via the endoscope) cannot be carried out without the property-owner's permission. Finally, if the cavities have been filled with insulation, it may require one or two bricks to be removed, to enable wall-tie corrosion to be confirmed. For further details, see **Wall-tie replacement** details, covered below.

Wall-tie replacement: This commonly used term in general Building is really referring to *wall-tie reinstatement* because the old, corroded wall-ties cannot be taken out and *replaced*, as such, they have to be renewed in nearby wall-positions with improved, modern ties, *before* having their rust-expanded ends treated and isolated. And as such work carries a liability of affecting the structural stability of a building already weakened, it should be regarded as a specialist operation. HELIFIX (www.helifix.co.uk) are such specialists, who can take over the whole operation from start to finish (survey to insurance cover, if required). They also hold a list of recommended installers, trained to use their products, which seem to rank high amongst the leaders in this field. Such products include the popular Helifix DryFix Wall Ties (a mechanical-pinning and remedial tying system) that requires no resin, grout or mechanical expansion. They are simply power-driven into both leaves/skins via a small pilot hole, regardless of the cavities being filled or partly-filled with insulation. In fact, the tie-insertions (*through* the insulation) should create less *cold-bridging* (via their smaller cross-sectional area) than the usually larger cross-sectional area of the old metal wall-ties being replaced. Additionally, in this company's range of products, there are the Helifix Resin Ties (ResiTies), or the Helifix Cem/grout Ties (CemTies), that become anchored into place when their resin (or grout) sets. But one Helifix-acknowledged downside of these types of ties is that they are not suitable where cavity-wall insulation has been installed. There are also Helifix Mechanical Expander (TorkFix) Replacement Ties, that consist of a threaded bar with sleeves, which are pushed through pre-drilled holes and then mechanically turned; this action expands the sleeves and creates a secure connection in the inner- and outer-leaves of walls built with relatively soft building-materials – such as lightweight, foam-aerated building blocks, etc.

Regarding the important business of finding and isolating/nullifying the corrosion of the outer-ends of old wall-ties, a wall-tie replacement specialist in Eastbourne, East Sussex, reckoned that his company's practice is to find the outer ends of the wall-ties via a metal detector, carefully remove a small area of mortar or brick/stone around each one and fill it carefully with expanding foam from a sealant gun. Then the recessed foam is eventually made good with a compatible mortar. My thanks to him are given in the Acknowledgements (enquiries@poultonremedialservices.co.uk).

Wall-tingle: When bricklayers lay a row of bricks, they usually work to a taut string-line that depicts the upper, face-side-arris alignment, related to the horizontal-levelled alignment of each course, as they bed and tap each brick into position. However, no matter how taut the line is, over certain lengths of wall being built, the bricklayer's line will sag. And, to remedy this, a single brick is carefully laid to the gauged height in the mid-area of the wall and is used to support the sagging line via a so-called **tingle**. This line-support, in its simplest form, might be a folded piece of cardboard, encasing the line and positioned (laid) on the central, gauged-height brick. After positioning the tingle to hold the line to the brick's arris, a holding-down brick is then laid edgewise on top to hold the tingle.

Warm-air central heating: The most popular of these systems used in the UK in the mid- to late-20th century, seems to have been the so-called forced-circulation system, where warm air is blown via a fan through a system of ducting concealed below the finished surface of the ground-floor. From the heater, the metal duct first enters a mani-fold, from which a number of duct-runs fan out to supply floor-connected grilles in the various rooms. However, with the present-day trend for overlay floors in a variety of quite expensive materials, the intrusion of metal grilles in any area of such floors might be unacceptable.

Warm-deck flat roofs: *Figure 80*:   The main feature with such roofs is that the insulation material is placed above the structural deck in the form of rigid insulation slabs, such as

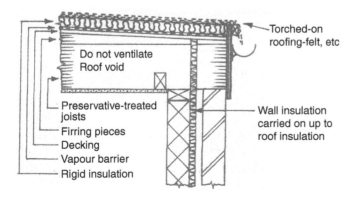

Figure 80 Warm-Deck flat roof, showing the eaves' detail

Thermazone polyurethane foam roofboards. These particular boards are sized 600mm x 1200mm and are 50mm or 80mm thick. Their edges are rebated for interlocking (to avoid cold bridging) and they have bitumen glass-fibre facings on each side. This rigid insulation must be bonded or fixed with mechanical fastenings to a layer of felt vapour-barrier, which itself must be fully bonded to a moisture-resistant sheet-decking material. The decking is fixed to tapered-timber firring pieces, which have been fixed to the roof/ceiling joists. To avoid cold bridging around the perimeter of the roof, the cavity insulation in the walls should be carried on up to meet the warm-deck insulation. In this type of construction, the roof-void must not be ventilated. All roof timbers should be preservative-treated and the final roof covering is recommended to be of rubber EDPM (ethylene propylene diene monomer), or GRP (glass-reinforced plastic).

Warranties on newly-built houses: The **NHBC** (National House Builders Control) warranty has been in existence for many years. The scheme has its own inspectors who visit new-build sites during the usual critical stages of the building process, and finally a signed warranty is issued, covering the quality of the work for a period of years. Local Authority Building Control (**LABC**) or Local Authority Building Partnership (**LABP**) departments now compete with the NHBC for this lucrative site-inspection service – and the issuing of fixed term warranties – by using their own Building Inspectors, or (nowadays) self-employed, nominated inspectors – in partnership with the local authority – to do the work.

   Apart from the length of a warranty (which is thought to last ten years), to my mind, the fact that a property was built under critical-inspection in the vital stages of construction, outweighs the fact that a warranty may have expired upon resale of the property.

Wash-hand basin/wash-basin: This self-explanatory facility, either wall-hung, housed in a so-called vanity unit, with a cupboard below, or – in recent times – incorporated into the top of a WC, has also joined the list of simplified, one-syllable Americanisms in fairly recent times by being wrongly called a **sink**.

Waste pipe: A 40mm/1½ in. diameter pipe, usually of plastic nowadays, with solvent-weld joints (to EN 1455, BS 5255), which is used to carry soiled water away from a sink, a wash-hand basin or a bath, via S-traps or P-traps with removable screw-caps for cleaning.

Water bars: See *Figure 35* on page 84: External doorframes usually have hardwood sills – sometimes referred to as thresholds. Traditionally, water bars, with a sectional size of 25mm/1 in., ran along a groove in the top of the sill, protruding by 12mm/½ in. to form a water check/draught excluder. These bars were made of brass or galvanized-steel, but nowadays, the bar is available in grey nylon or brown rigid-plastic with a flexible face-side strip, which acts as a draught seal against the rebated, bottom-edge of the door.

Water-based paints used externally: These modern paints are water-soluble, but become water-resistant when they dry. As a binding agent, different ingredients are used in the paint's manufacturing stage. These include acrylic, vinyl, alkyd or polyvinyl acetate (PVA), giving water-based paints many advantages over traditional oil-based paints. For example, they emit far fewer health- and environment-damaging VOCs (volatile organic compounds) than oil-based paints – and the painting equipment used (brushes, containers, etc.) can be washed out more easily in water. Water-based *paint* (not *emulsion*) is also more compatible with the natural, moisture-content characteristics and movements of timber. Thereby, it is reckoned to last longer than oil-based paints, meaning that maintenance periods can be at least doubled from the traditionally recommended four years to eight or ten years.

Water closet (WC):   Historically, this term meant a small room which contained a waste-piped toilet pan and a crude flushing arrangement. However, the term WC is now used – at least in written references – to refer to the toilet pan itself, with an attached flushing cistern. Personal references to it seem to be toilet, or loo, etc. But what must be mentioned (from a *water-rates*-saving and an environmental-viewpoint) is that the traditional cistern (operated by a lever handle, or a pull-chain), which is still very much in existence, does not apparently waste water as much as a modern, push-button cistern. The lever-handled, or pull-chain cistern uses siphonic action, via a ball-valve or a control-float valve to operate a diaphragm-plunge washer, which rarely needs maintenance – as opposed to the more modern cistern that uses a valved flushing cistern operated by a dual-flush action push-button – which seems to need cleaning out periodically, to eliminate the unhealthy-looking build-up of dark, toilet-pan staining that emanates from the flushing holes under the pan's rim.

Note that in a BBC radio programme 'Costing the Earth' in October 2020, the reporter's research uncovered the fact that base-valved flushing cisterns with dual-flush push-buttons tended to waste a considerable amount of water over a period of time. This was because the cistern's water leaks away unknowingly through minute debris particles in the water affecting the seating of the valve at the cistern's base. The programme also questioned the logic of the dual push-button design, regarding which one should be used to serve our dual needs? Two out of three members of the public did not know. Of course, one could always experiment to find out which part of the button provided the *lesser* or *greater* release of water. The former being usually required for the more frequent bodily function.

Water hammer: This is a banging noise which can occur in high-pressure water pipes – and the noise is usually caused by one of two likely reasons: **1)** the noises could be related to a loose *jumper* in the mains stopcock, which jumps too quickly onto its seating or it might be caused by loose pipes that have been fixed with too few pipe-clips. Such poorly clipped pipes are known to vibrate and bang against each other, after a running tap has been closed too quickly; **2)** the second likely reason for noises in pipework might be to do with a float-operated ball-valve in a cold-water storage cistern, which is being unduly disturbed by movement of the cistern's surface water, causing the ball-float to oscillate up and down. This, in turn, opens and recloses the ball-valve too frequently – causing an audible noise.

Water seal: See *Figure 52* on page 142: This refers to the *safety-seal* created in the S-and-P-configured design of S-traps and P-traps connected to sinks, wash-hand basins and WCs, etc. They stop odours rising up from the sewers and from intercepting traps in traditional inspection chambers (also known as *manholes*). The water seal is created by the downward-loops of S-and-P-shaped pipes (when such shapes are rotated by 90^0 to the right), by water finding its own level above the invert of the outward bend, which leaves it above the inverse shape of the lower, inward bend, thus creating an odour-preventing water seal.

Water-storage reservoirs: Domestically, these range from a simple water-butt, a range of different shaped plastic vessels submerged in the ground, to a large underground storage tank, with an electric pump, that all collect rainwater for garden use.

Water tables: In building terminology, this term is a reference to the surface of underground water, below which fissures and pores in the strata are saturated with water. Under high ground, the water table is usually at a greater depth below the ground surface than below inverted contours (such as valleys), where rock formations are more permeable. Water flows through them easier, therefore water tables are lower. But as water sinks into the ground, the water table rises. The height of a water table – and its likely penetrating pressure on any underground structures such as subterranean basements and cellars – have to be considered carefully at the design stage.

And (on site), any specified structural treatment, such as three-layered asphalt tanking, sandwiched between a double-thickness concrete oversite and between double-leaf walls, must be carefully supervised. For example, when the inner walls are being built against the vertical, asphalt surfaces, they should be *tight* up against the asphalt. And this can only be done by either progressively *buttering* the inner faces of the bricks as they are being laid on the bedding mortar, or by laying a course of bricks and then infilling the unavoidable gaps against the asphalt. In my experience, unmonitored bricklayers do not like doing either of these operations (both of which are tedious and time-consuming). But such ill-considered omissions, can cause hydrostatic water pressures to fracture the asphalt surface.

Wayleave: Legal, documented permission to pass over another person's land or property, which may also include the right to lay pipes or cables, etc.

WC: See **Water closet (WC)**

Weather bars: See **Water bars**

Weatherboarding: This description usually refers to tanalized or protimized (preservative-treated) boards used for the exterior cladding of sheds, garages and fences, etc. They vary in design, but *featheredge boarding* is commonly used on timber-framed garages and sheds, with a weathered-overlap of at least 13mm/½ in., and is widely used on *post-and-arris-rail fencing*, with a similar overlap to counteract shrinkage. Other board designs usually include rebated or tongue-and-grooved edges and are referred to as *shiplap boarding.*

Weather boards: See *Figure 35*: These traditional appendages, still in use nowadays, are usually rebated to produce a projecting tongue and are fixed on the outer-face of exterior doors, via a horizontal groove in the door, close to the door's bottom edge. They commonly have a top-sloped (weathered) surface of about 40^0, or, more tradition-ally, a cyma-recta moulded surface.

Weathered-slopes: See **Coping(s)**

Weathered, struck-and-cut pointing: See **Pointing**

Weathering/weathered timber: The pitted and chemical break-up (but not technically decay) of timber surfaces – especially end-grain – exposed to rain and sun over a period of years.

Weather-resisting and **weatherproof glues:**   Most exterior adhesives/glues are rarely weather*proof*, unless being used by high-frequency equipment in factory conditions. The most-used weather-resisting glues (regardless of their manufacturers' names) seems to be **1)** *PVA* (polyvinyl acetate). This is a resin type of glue in the form of a thick, white liquid. It is applied directly from squeezable plastic containers and/or by brush; **2)** *Synthetic resin glue*, such as Cascamite (in powder form), which is chemically referred to as urea formaldehyde, a two-part glue that is activated when mixed with water; **3)** *Phenol formaldehyde* adhesives that have a high water resistance when used for exterior timber-jointing; and **4)** *Resorcinol formaldehyde* adhesives that have excellent water-resistance on exterior use.

Note that a typical specification for engineered-timber structures, such as plywood web beams and gluelam-timber portal frames (the latter often used externally) have been seen specified as having used 'fully weatherproof resorcinol-phenol formalde-hyde, Aerodux 185, supplied by Messrs CIBA (ARL) Ltd to comply with BS 1204 WPB and GF'.

Weep holes: This term is used to describe raked-out perp-joints (perpendicular mortar-joints) at the base of the outer-face brickwork of double-skin (double-leaf) cavity walls to allow the escape of any moisture that may have entered into the cavity. These weep holes were also formed by the bricklayer placing patented plastic-sleeved weep holes in the selected perp-joints as the bricks were being laid. Sadly, they seem to be mostly omitted nowadays.

Wet cellar: This description is sometimes used as a *fait accompli* statement when being reported upon after a Building survey. It is particularly applied to early-Georgian period houses with a cellar. Occasionally, such subterranean rooms turn out to be not much more than useful inspection chambers, seemingly built to enable a periodic appraisal of the saturated brick walls to be carried out. Such walls can often be seen to be weeping with lateral ground-moisture penetration – below a possible non-existent damp-proof course – and from a high water-table on their exterior ground-abutment. In such cellars, it is not unusual to walk through puddles to check whether a sump-pump has been inserted in the cave-like floor.

However, the structural condition of the cellar/foundation walls needs to be carefully checked for erosion of the mortar joints and/or spalling/crumbling of the brickwork. And, remedially, a French drain around the outer-face of the basement walls could be considered.

Wet rot: This benign timber-destroying disease, unlike *cellar rot*, or *dry rot*, comes under the classification of being a *white rot* wood-destroying fungus, and is known for its destruction of poorly maintained exterior woodwork, such as wooden windows, doors, fascia- and barge-boards, fences, etc. However, it is not malignant and the rotting process – triggered by alternating wet-and-dry conditions of exterior exposure – will cease *if* the exposure to moisture is curtailed. This can be achieved by repairing and protecting the damaged items with suitable proprietary wood-hardeners, fillers, preservatives and/ or paint, etc. Nowadays, pressure-impregnated, preservative treatment of timber has considerably reduced incidents of wet-rot timber decay – but there are still a considerable number of ill-maintained old-build properties around with either seriously-degraded timber sash- or casement-windows, fascia- and barge-boards, etc., which, upon closer inspection, might confirm that they are beyond repair. Note that although wet rot is botanically classified as a *white rot*, contradictorily, the colour of the affected, rotten timber, is very similar to *dry rot*, i.e., brown, but, if the timber is saturated, black streaks may be present. Another similarity is the severe destruction of the cellular tissue.

Whitewood: See **Redwood and whitewood**

Wind, winding: These timber-terms (pronounced *wined, wineding*) are the equivalent of Twist and Twisting. The term 'in wind' means twisted and, therefore, 'out of wind' means not twisted.

Wind-bracing: See **Restraint-strap additions**

Winding: See Wind, winding

Winding steps: See **Tapered treads (steps)**

Window boards: As windows are usually positioned near the outer edges of the openings in a wall, the remaining inner portions at the sides (the **reveals**) and at the head (the **soffit**) and at their horizontal bases (*inner sills*), are relatively wide and need to be covered with a *finishing* material. So, the soffits and the reveals are usually plastered or dry-lined, but the base (the inner sill), because it is used like a shelf and/or is more exposed to wear-and-tear, is usually covered with a *window-sill board*, more commonly referred to as a *window board*. It may be of hardwood or softwood timber, usually of 28mm/1⅛ in. finished thickness, with a nosing projection; or MDF (medium density fibreboard) of 18mm/¾ in. thickness. Both materials should have bullnosed or rounded outer-edges. Other options for the inner-sills include quarry- or ceramic-tiles, stone or slate, etc.

Window-sash locks: There are two types of sash-fasteners for traditional sliding sashes, one shaped like an ear that receives a side-levered arm to pull the central meeting-rails together – the other, more modern, has a pivoting, threaded bolt with a wing-nut attachment that folds to the side when open, and is raised, turned and dropped into a receiving-staple on the opposite meeting-rails, when being locked by tightening the wing-nut.

   Although these effectively lock the sliding-sash windows together, they have been documented as failing in a burglary and ought to be backed up with patented *sash-window retractable stops*, of which there are a few types available at ironmongery outlets.

Window shutters: See **Folding shutters**

Withe: See **Wagtail** or **Chimney stacks**

Wooden pendulum: See **Wagtail**

Wooden rolls: These are of softwood and are used on flat roofs to be covered with sheet-lead. Their finished size is specified by the Lead Development Association as being of 45mm x 45mm (1¾ in. x 1¾ in.), with a semi-circular top half and sloping sides from the springing line down to a centrally-flat base of 25mm/1 in. width. The reason for the provision of wooden rolls is to effectively weather the joining of the side-edges/ joints of the relatively narrow rolls of lead, via the sides of the lead being upturned and *dressed* over the roll on one side, onto the lessened, lead upstand on its opposite side. Also, the provision of lead-roll bays helps to control the rate of expansion and thermal movement of sheet lead, which is sometimes referred to as *creeping*, because of its insidious movement over a period of time.

Wooden shingles: As used for roof-cladding on pitched roofs in some parts of Europe and England since the 15th century – and still used to some extent in a few countries. They can be thought of as wooden slates, as their bonding and overlapping arrangement is very similar. Shingles are manufactured from various species of durable timber, such as western red cedar, certain species of oak and red cypress, etc. Like any other roof cladding, these would need to be under-clad with a water-resistant membrane.

Woodscrews: Screws have changed considerably in the last decade and we now have twin thread, deep-cut thread and steeper thread angles extending along the entire shank to provide a better grip and speedier driving performance. There are also now at least seven variations in head-design, ranging from the traditional straight-slotted head (which I cannot imagine being used nowadays – certainly not professionally) to the Torx- or T Star-slotted head. With the revolutionary change about three decades ago from hand screwdrivers to electric- and cordless drill/drivers, the development in slotted-head types was necessary to create a more secure, non-slip driving attachment. Improved slotted heads such as Uni-Screw and Torx/T-Star promote non-wavering driving alignment, virtually eliminating so-called *camout* situations.

Like nails, screws also come in collated strips for use in mains-powered, automatic-feed screwdrivers. The screws for these drivers have a bugle-shaped head, which is an innovative countersunk shape allowing them to sit at the correct depth in plasterboard without tearing the paper. These auto-feed drivers have been marketed particularly for dry-wall operations – fixing plasterboard to lightweight steel or timber-stud partitions – but can be used on other repetitive screwing operations, such as fixing sheet material to wood floors prior to floor-tiling; or sheet material to flat roofs prior to felt overlay or GRP roofing. For these operations, collated floorboard-screws, with ordinary countersunk heads, are available.

Woodturning: Making circular wooden items – such as spindles – with a woodturning lathe.

Woodworking adhesives: See **Weather-resisting** and **weatherproof glues**

Woodworm: This widely-adopted reference to the *common furniture beetle*, of the *anobium punctatum* genus, refers to the larvae of the beetles that bore a complexity of destructive, interconnecting channels (bore holes) within the body of hardwood items and softwood building-timbers. The latter include roof members, floor joists, floor boards, the underside of stairs, etc. Usually, the low number of small-diameter (1 to 1.5mm Ø) flight holes seen on the surface of timbers (caused by the larvae, after they have had their feed, developed into beetles, bored their way out and taken flight) belies the extent of the internal, honeycombed destruction. Occasionally, the bore-dust (*frass*) can be discovered near the flight holes in undisturbed roof-voids and exposed underfloor-troughs; I have occasionally found it held captive in networks of cobwebs. It resembles saw-dust and is ellipsoidal in shape.

These extremely destructive insects were allegedly brought into the UK by people bringing pieces of furniture over from Europe; hence the reference to this in their name. Oddly enough, though, they seem not to be found so commonly in furniture nowadays.

In situ treatment of roof-void timbers and floor joists, etc., with insecticides, is possible, but because of the inherent dangers of individuals using such products in relation to the Health and Safety at Work Act (HASAWA) and the Control of Substances Hazardous to Health (COSHH), such treatment is best left to reputable specialist-companies. Their work should also provide a written guarantee of non-recurrence for a stipulated period of years.

Worktops/counters: These kitchen work surfaces are generally referred to as *worktops* in the UK, but the Americans seem to refer to them as *counters*, so the term is somewhat blurred. Nowadays, however, with all the hi-tech kitchens and granite, quartz/polymer resin and Corian® worktops, etc. being marketed, the fitting and fixing of them are now very much a specialist's job.

Work triangle: *Figure 81*:   In the early 1950s, researchers from Cornell University in New York, USA, conceived the idea of *the work triangle* as being the geometry determined by the critical positions of the sink, the refrigerator and the cooker. These were identified as the three main centres in a kitchen involving traipsing backwards and forwards during the cooking task. Study revealed that an imaginary line joining the sink, fridge and cooker should ideally form a triangle measuring no more than 20

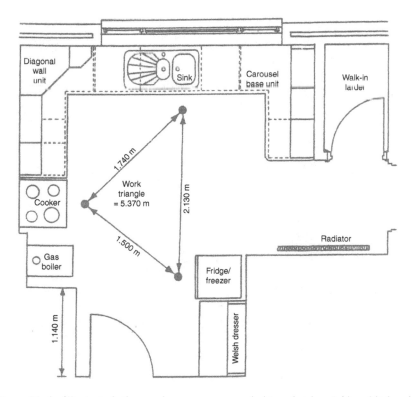

Figure 81 As illustrated, the maximum recommended 'work triangle' in a kitchen is 6.1m/20ft

ft./6.1m. in total. It was also established that the distance between opposing worktop-edges, as in a narrow, so-called galley kitchen, should be no less than 4 ft./1.219m.

Wreathed handrailing: See *Figure 25* on page 64: **Continuous (wreathed) handrailing.**

Wreathed strings: See *Figure 26* on page 64: Basic staircases contain two inclined wide-boards-on-edge that house the ends of the steps – and these boards are called *string*s. The board against the wall is referred to as the *inner-string* or *wall-string* and the board away from the wall (i.e., self-suspended) is called the *outer-string*. So, when a staircase is designed to fit around a curved wall (or up against a normal, straight wall, but with a curved outer-string), they are both referred to as *geometrical staircases*. In the first case, both strings will require to be wreathed (as will the handrailing), and in the second case, there will be a normal, straight string, but a wreathed handrail above the outer, wreathed string. Being wreathed means being curved in plan-view and spiral-shaped (twisted and coiled/helical) in the elevational-view.

Wrot timber: This is an archaic term for *wrought* timber, meaning *surfaced* timber, which might still be used by quantity surveyors. Mainly, in the industry, the term used is *planed* or *prepared* timber.

Wrought timber: See **Wrot timber**

W truss: See **Fink** or **W truss**

X

X-ended wall-tie bars: See **Wall-tie bars**

Y

Y-ended wall-tie bars: See **Wall-tie bars**

Z

Zinc: A bluish-white metal, traditionally used as thin-gauged sheet material for roofing areas such as dormer flat-roofs and side-cheeks, soakers and flashings, etc. For longevity, this material is regarded as inferior to heavy-gauge sheet lead, copper, modern torch-on felts and sheet-rubber (EDPM).

9 781032 253916